W9-BLY-543

DECODING DARKNESS

DECODING DARKNESS

The Search for the Genetic Causes of Alzheimer's Disease

RUDOLPH E. TANZI

ANN B. PARSON

PERSEUS PUBLISHING

Cambridge, Massachusetts

Rudolph Tanzi is a principal scientific founder of Prana Biotechnology, Ltd.; Genoplex, Inc.; and Neurogenetics, Inc. Dr. Tanzi has equity in all three companies as well as the publicly traded companies Elan Pharmaceuticals and Bristol-Myers Squibb.

A CIP record for this book is available from the Library of Congress.

ISBN: 0-7382-0195-2

Perseus Books is a member of the Perseus Books Group.

Find Perseus Publishing on the World Wide Web at http://www.perseuspublishing.com

Text design by Heather Hutchison
Set in 11-point Goudy by the Perseus Books Group

1 2 3 4 5 6 7 8 9 10—03 02 01 00
First printing, September 2000

For the Noonans and all others confronted by this disease,
and for
Ann M. & Rudolph A. Tanzi, and
Katharine & Christopher Rodgers

Contents

Illustrations

Acknowledgments

The idea for this book occurred on Boston's Storrow Drive on November 7, 1995. That day's horoscope in the *Boston Globe* seemed to indicate a go. Its advice: "Eliminate the word 'impossible' from your vocabulary." Keeping that word at bay ever since, a wide circle of people have fortified and inspired this book project each step of the way. Very certainly, it's impossible to thank you enough.

In special, we are deeply grateful to Joy Glenner, for her shared remembrances of her husband and her enthusiastic support; to Cai'ne Wong, for his scrupulous descriptions of the science in George Glenner's lab; to Jim Gusella, for his considerable guidance and fact-checking, especially as pertains to the Huntington's work; to Dennis Selkoe, for his close reading of several chapters, his helpful corrections, and his contribution of so many key details; to Bill Comer, for innumerable wise insights into the pharmaceutical world and for being a savvy mentor; to Robert Terry, for laying out the historical backdrop and for also being a savvy mentor; to Wilma Wasco, John Hardy, Mike Mullan, Jerry Schellenberg, Christine Van Broeckhoven, Dale Schenk, and John Breitner for their extensive input and review of certain sections; to Paul Raia, for his open-door counsel; to Jean-Paul Vonsattel, for his in-depth tutelage in neuropathology; to Kathleen Ottina, for her patient explanations; and to the triumvirate partners—Steve Wagner and Sam Sisodia—each of whom lent large. We are indebted as well to Carmela Abraham and Dora Kovacs, for their insights and for permitting us to use their photographs. Posthumous tribute goes to Henry Wisniewski who left us with many valuable points of reference, as well as the mandate to tell some good stories.

The Noonan family gave this book the invaluable reality of what an at-risk family is faced with. To Julie, Pat, Malcolm, Eryc, Fran, and the others, including spouses, our great appreciation for sharing your lives and imparting your strength as a family to the rest of us.

If we have managed to pin down the details, we are indebted to still others: Mary Anne Anderson, Larry Altstiel, Mark Baxter, Merrill Benson, Katherine Bick, Deborah Blacker, Danny Chun, Alan Cohen, Linda Cork, Peter Davies, Kevin Felsenstein, Blas Frangione, Robert Katzman, Kevin Kinsella, Virginia Lee, Ivan Lieberburg, Rachael Neve, Toni Paladino, Huntington Potter, Richard Roberts, Kathleen Sweadner, John Trojanowski, Paul Watkins, Bruce Yankner, former SIBIA-ites, and all others who supplied expertise.

Our gratitude goes as well to the Special Collections staff at the University of California, San Diego, for making available George Glenner's papers; to the Oral History Program, Department of Special Collections, University of California, Los Angeles, for granting permission to draw upon the oral history interview it conducted with Rudy for the Pew Scholars in the Biomedical Sciences Oral History and Archives Project of the Pew Charitable Trusts; and to the helpful librarians at Harvard's Countway Library of Medicine and Boston Public Library.

Members of Mass General's Genetics and Aging Unit deserve medals not only for their ready assistance, but for being so accommodating of this project. To Wilma Wasco, Dora Kovacs, Ashley Bush, Donna Romano, Tae-Wan Kim, Deirdre Couture, and Denise McDougall—tremendous thanks to each of you.

Fortunately, another member of the above clan is Rob Moir, a protein chemist whose fine talent for computer graphics lends so much to these pages.

That this book actually took three-dimensional form is due to three ardent bookmakers. Thank you, Jeffrey Robbins, for your excitement over this project and for taking it on. Thank you, Doe Coover, for being such a discerning and nurturing agent. And thank you, Amanda Cook, for your wise decisions about content and for being such a first-class editor. The book's production team, headed by Marco Pavia, made it all come together: large thanks for all your efforts.

When the word *impossible* especially threatened, family and close friends vanquished it. Please know—each one of you—how much your steadfastness has meant.

RET & ABP

Introduction

Few real nightmares on earth compare to the terror wrought by Alzheimer's disease. That this fatal brain disorder not only annihilates a person's mind, but starts doing so years before it takes their life is surely its most insidious aspect. Its initial symptoms of forgetfulness and personality changes lie so close to normalcy that they typically go unnoticed; and, once noticed, too long unexplained. As the victim's grasp further slides, it can bring nothing but tormenting confusion for both the patient and those close to him. What can be worse than watching someone you love cognitively flailing, until eventually they no longer recognize faces, surroundings, or even themselves?

More razor-sharp than Alzheimer's physical distress is this emotional pain felt by patient and helpless bystanders. "No one not saddled with it can understand it, not even my best friend," says Julie Noonan-Lawson. Along with her four sisters and five brothers, Julie watched their mother Julia Tatro Noonan succumb to a rare form of Alzheimer's that strikes in middle age and is passed down to 50 percent, on average, of offspring. Consequently, all ten of Julia's children, who currently span the ages of thirty-six to fifty-eight, bear the burden of being genetically at risk.

Framed in their recollections of their mother is how much she loved to sing. Growing up, they would cram into the family station wagon on hot summer days, swimsuits in tow, and led by Julia's strong lilt sing one song after another full throttle all the way to Manomet Beach on Massachusetts' South Shore. When, in her early forties, Julia inexplicably began singing less and lapsing into depressed moods, her children followed her into a mire of anxiety, trouble, and hurt. Her lost song was their lost song.

The Noonans' response to the disease has been to not take it sitting down. So admirably, they and hundreds of other at-risk families have made invaluable contributions to research, helping the thousands of us who make up the Alzheimer's scientific community to try to extrapolate the disorder's molecular roots. Without their aid, the following account of the inestimable progress we've made in a remarkably short time wouldn't exist for the telling. Faced with the prospect of the disease's bull run through generations of their large family, the Noonans made their DNA available to investigators at Massachusetts General Hospital in Boston. It's primarily under that roof that I've been involved in researching Alzheimer's causes since the early-1980s, currently as the director of the Genetics and Aging Unit. From this vantage point emerges the following story. Although words can't fully describe the fears and losses an Alzheimer family is up against, some small sense of the Noonans' ordeal appears between these chapters.

When a doctor recognized Julia's illness in 1967, its brand of dementia was thought to be confined to middle age. It wasn't long before the true boundaries of Alzheimer's emerged. Researchers realized that its classic lesions—the microscopic *amyloid plaques* and *neurofibrillary tangles* that overrun brain tissue—also appeared in elderly people who suffered from senile dementia, and with alarming frequency. Since then, the disease has been perceived in two guises. The extremely rare type, which first manifests in people under sixty, is known as *early-onset* Alzheimer's. This is what plagues the Noonans. When the very same plaque-and-tangle pathology descends on people sixty or older, it is referred to as *late-onset.* So common is this late variety that in this country it afflicts 20 percent of people age seventy-five to eighty-four, and reportedly over 40 percent of those eighty-five and over. No other neurodegenerative disease takes so many lives.

Although its extensive toll stood revealed, by the 1980s Alzheimer's was still considered a backwater disease, one that didn't attract much research attention. Those who studied it were coming round to believing that its early form was inherited, and therefore the result of a genetic error. Late-onset cases, on the other hand, were thought to be caused by environmental stressors, not mutations in genes.

Generally, Alzheimer's grip on older people cooled the interest of researchers. The unfortunate sentiment was, why work on a disease whose victims were close to the end anyway? More to the point, the technology available for exploring the mysterious complex of the human brain was limited. Even today, despite two decades of tremendous strides in neuroscience, a central irony persists. The human brain has festooned Earth with amazing objects of its own making—high-speed computers, Boeing jets, buildings that touch the sky, powerful medicines—even elaborate scanners such as MRI (magnetic resonance imaging) and CAT (computerized axial tomography) that further our attempt to understand the brain as well as we do the heart and lung. Yet as Stephen Vincent, a neuroscientist at McLean Hospital outside Boston, so aptly puts it, "Our brain is having a devilish time figuring itself out. It remains totally ignorant of how it works as a unified organ to make us so uniquely human." Even a slug's brain, and how it enables a slug to be a slug, eludes scientists, notes Vincent. If the trappings of a healthy brain are hard to decipher, imagine the murky picture a diseased brain presents.

In the early 1980s, the genetic revolution's fantastic inroads into the biological world's two crucial elements—*genes* and *proteins*—began making possible a more introspective investigation of human diseases, including those that rob the brain.

Genes, composed of the molecule DNA (deoxyribonucleic acid), are the basis of all inherited traits in every living organism on Earth. Each gene is a linear chemical script, a special cryptogram made up of DNA's four nucleotide bases—adenine (A), guanine (G), thymine (T), and cytosine (C). And each is part and parcel of a far longer DNA filament—a chromosome. So genes simply amount to codes of inherited information that are intermittently spaced out on chromosomes. A human has twenty-three pairs of chromosomes—one in each pair inherited from mom, one from dad—which present us with two copies of roughly 100,000 genes. This full complement is found inside the nucleus of most every cell. Although we each inherit two copies of every type gene, there are, in fact, several different versions of each gene floating around in the population's overall gene pool, any of which may be inherited. Versions of the same gene can differ by just a few bases.

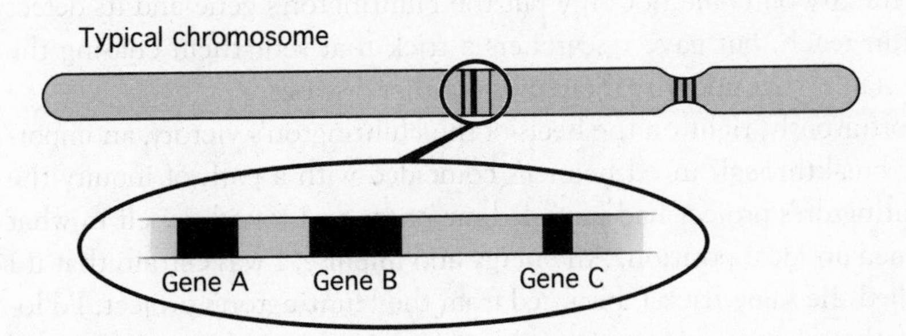

FIGURE I.1 Genes interspersed along a chromosome.
Illustration: Robert D. Moir

Since genes and proteins are so central to the story that follows, pretend, as I often do, that they're characters in a play. Like kings, genes more or less stay seated on their thrones inside the nuclei of cells. But most every gene imparts its crucial code for the making of a protein, proteins being the slaves of the biological world. Once encoded by genes, proteins take care of nearly every function necessary to an organism's survival. In humans, this work ranges from building tissues for organs, to making enzymes that exert chemical changes for food digestion and other necessities, to practically everything else that keeps the body's trillions of cells happy and ticking.

Most important, genes provide the instructions for proteins. The structure of a gene and its corresponding protein therefore are interrelated. If a gene happens to contain a serious flaw, or mutation, in its lineup, the making of its protein can be aborted. Sometimes, however, a defective protein survives, and its faulty amino-acid structure mars its normal activity. Depending on how much it deviates, this can lead to a specific disease.

In 1980, by a stroke of considerable luck and timing, I happened to join a team that went on to demonstrate a novel technique for finding disease genes and their proteins. Back then, very few human genes had been identified. They lay in the blackness of a cell's nucleus, absolutely tiny and inscrutable. For a twenty-one-year-old who was enthralled with genes to begin with, to land on the doorstep of this undertaking (headed by James Gusella at Mass General) was beyond any dream I had for a career. The experiment involved trying to locate the gene defect that underlies Huntington's disease, a neurodegenerative disease, which like Alzheimer's probably has damaged humanity since Cicero's day. The extraordinary outcome not only put the Huntington's gene and its defect within reach, but gave researchers a trick that sent them chasing the gene faults that underlie thousands of other diseases.

Fortuitously, right on the heels of the Huntington's victory, an important breakthrough in Alzheimer's coincided with a path of inquiry the Huntington's project had funneled me onto. So I found myself in what seemed an ideal position. All energy and impulse, I was certain that if I applied the same trick I'd learned from the Huntington's project, I'd locate the mutant gene responsible for the rare, early-onset form of Alzheimer's—a first step that might lead to a treatment for both early and late forms of the disease.

Little did I or others embarking on the same investigation realize the morass we were heading into. The total detective work into Alzheimer's since then, which has been carried out by a swelling universe of scientists from different disciplines and continents, has been anything but straightforward. Yet in less than two decades we've gone from knowing little about what causes Alzheimer's dementia—which is the outward manifestation of severe brain-cell degeneration—to knowing what goes wrong on several fronts. Exactly *how* it happens still eludes us nevertheless.

The thread for this book is the evolution, clue by clue, of one particular hypothesis that has gained widespread support. It poses such a believable explanation that not only have hundreds of us sold our souls to it, but nearly every large pharmaceutical company and countless biotech companies are in the process of crafting a new generation of drugs based on its general principle. As the year 2000 begins, the first of these treatments, having shown promise in mice, are just beginning the ultimate challenge—testing in humans. Even those of us who have contributed to the backbone of this hypothesis are apt to blink in amazement. Not long ago the idea of curbing cognitive loss seemed as futile as trying to stop a spring tide from going out. To now be at the point of testing drugs woven from this credible theory seems nothing short of a miracle.

Anchored in molecular genetics, the hypothesis in the spotlight has to do with amyloid plaques, the microscopic deposits of protein that abnormally inundate the brain tissue of Alzheimer victims. The premise behind the *amyloid hypothesis* is that these deposits—or their subunit fibril—are directly to blame for the brutal wasting of neurons. This theory represents the dominant viewpoint, but keep in mind it's only a theory. You might think because it's backed by such heavy betting there would be general agreement over it, but this is far from the case. Numerous experts from inside and outside the field still view any drug prototypes arising from it as a misinformed long shot.

While the amyloid hypothesis may be the most popular explanation for Alzheimer's, plenty of other theories and resulting drug designs still undergoing research and development have come into existence. Although *Decoding Darkness* examines these potential therapies in less depth, it endeavors to show that this rich range of prospects represents the true prize. There have been times when our field has been criticized for crowing prematurely about how close we are to effective drugs. Yet today, derailing this scourge of an illness is no longer a dream but a distinct possibility.

Finally, the Alzheimer research field has gained a certain notoriety for being a fiery, truculent, and driven lot—moments of which we've allowed in here. What isn't talked about and written about is the considerable camaraderie that, through the years, has grown up among so many of us. While we may not always see eye to eye, we are stuck together like Velcro against a much bigger foe. Hats off to all those many colleagues, and may we soon know the answer.

DECODING DARKNESS

The proud father relayed the news over a candlestick-style telephone to family and friends in Agawam, Massachusetts, and beyond. "It's twins! God Bless! Lil has given birth to identical twin girls!" Before long, tiny Julia and Agnes Tatro were taking their first tottering steps and trailing each other into mischief. They were the spitting image of each other, from their wispy flaxen locks to their knobby knees—except for the deep dimple that Julia wore on her left cheek and Agnes wore on her right. It was the 1920s, the days of new-spinning Victrolas and sporty Model Ts; of old-fangled doctors whose black bags contained, first and foremost, a stethoscope and vials of morphine. Medicine's reach was skin-deep; the source of most ills too complex to grasp. People knew about genes—in the case of identical twins, their identical set—but they were as distant as magma at the earth's core, and no one knew what they were made of or the good or bad extent of an individual's inherited array. And so, even as Julia and Agnes Tatro blossomed into pretty, spunky girls who reached out to life, no one suspected what lurked beneath the surface. And it gave no sign of itself. It inched along its harmful course in too quiet a way, just as, in all likelihood, it had been doing even before their mother knew she was pregnant.

— *one* —

Cleave, Zap, Blot, Probe

It came like a lightning flash, like knowledge from the gods.
—Edward O. Wilson, "Naturalist," on Watson-Crick 1953 discovery of DNA's structure

At twenty-one and fresh out of college, I entrusted myself to the Taoist philosophy that the less you interfere in Nature's course, the more likely you will find your true path in life. This wisdom flowed from a slip of a book I'd discovered in high school—the *Tao Te Ching*. In retrospect, it would seem that giving myself up to "the way of things" succeeded, because that fall, out of the blue, an opportunity of a lifetime presented itself, one that introduced me to a spectacular new scientific method and later prompted my investigation into the genetic wrongs of Alzheimer's disease.

It was a cloudy September Saturday in 1980, and after the quiet of summer, Boston seemed energized by autumn's return. Beacon Hill's narrow streets were clogged with cars, its crooked-brick walks filled with residents and students who seemed all business. On the Charles River even the sailboats crossing the watery line between Boston and Cambridge flew forward at a clip. A few blocks east on Blossom Street, which curves behind Massachusetts General Hospital, members of the rock band Fantasy and I moved more like laden barges. Sleep-deprived and hungover from the previous night's fling, we nonetheless managed with an elevator's aid to move the band's musical equipment into the Flying Machine, the nightspot atop the Holiday Inn that attracted everyone from visiting Portuguese sailors to the occasional Brahmin.

Four months earlier, in May of 1980, the University of Rochester had sent me into the world with what I hoped would be sufficient padding—

bachelor's degrees in both history and microbiology. The one, Time Past, had filled me with an indelible impression of the patterns and trends that span recorded centuries. The other, Emergings of Time Future, had left me startled by the phoenix soaring out of the present—the molecular-genetics revolution. Biology's horizon was filled with elaborate possibilities far beyond the imaginings of such tour-de-force microbe hunters as Louis Pasteur, Robert Koch, and Paul Ehrlich.

In the course of my history studies, I'd devoured Thomas Kuhn's *The Structure of Scientific Revolutions* and taken away his valuable model. One set of beliefs ascends over time, then falls under the weight of a crisis, which inevitably ushers in yet another belief system that rises and similarly collapses, and so on, until there's a sense, as from a wave rolling forward, that you can extrapolate the nature of the next crisis and the new visions it will unfold. Now a scientist, I'm even more aware that the models we put our faith in are mostly wrong. Someday they will be as outmoded as the idea, imagined by Franz Mesmer in the eighteenth century, of how to relieve people of disease: Stand them across from healthy folk in a tub of water, have both groups grasp a long metal chain, and let the positive forces of animal magnetism flow from the healthy into the infirm, miraculously curing them. For scientific revolutions to take flight, current theories have to be questioned, the status quo disrupted. Since my years at Rochester, I've always wanted to induce the next crisis, inspire the next paradigm shift. This is the challenge of science—to shed dogma and get closer to the truth.

But scientific revolutions were the furthest thing from my mind that Saturday atop the Holiday Inn. I was in the throes of a postcollege existentialist crisis. Why did I exist? What *was* life? Living life as a bushy-haired, scruffy musician and playing keyboard once again with my musician friends from high school days seemed the best way to regain some perspective. When I was ten, my Uncle John had let me fold and unfold the huge red accordion he played in old-age centers around our hometown of Cranston, Rhode Island, and from then on I'd been glued to the keys of pianos, electric organs, and synthesizers. Blues, jazz, rock, punk, improv, some classical. One form fed another. I'd come to realize that when I played music on a daily basis—even on an informal basis, as I had throughout college—life was always better. When I didn't, disaster struck.

With college behind me, I filled in on keyboard for various friends' bands. Night after night, sometimes for seven nights in a row, we sang and gyrated in smoky bars and plush, mirrored clubs strewn between Boston's Kenmore Square and Providence's East Side. After the dreaded repacking of equipment, near dawn I commuted back to my mother's house in Cranston for a precious few hours' sleep. All in all, I wasn't seeing too much daylight. Although I sometimes fantasized about it, a career in music was unlikely. Ever since I accidentally was knocked unconscious on stage by the solid-body Stratocaster hurled by Ritchie Blackmore of Deep Purple fame, I had had second thoughts.

Fantasy's show at the Holiday Inn was serving as warm-up for Jan and Dean and their oldie-but-goodie surfing songs. The scent of fall in the air must have awoken a desire for more permanency in my life, because once we had things set up, on a whim I walked around the corner to check out the job postings in Mass General's Bulfinch building. One particular notice caught my eye: "Assistant needed for study addressing the genetics of neurological disease. Experience in genetic linkage and restriction enzyme digestion required. Contact James Gusella."

As a boy I'd viewed Massachusetts General Hospital (MGH) as a Mount Olympus peopled with white- and green-coated demigods who always had some urgent place to go. My parents ran a medical transcription service, and they often took me and my twin sister Anne with them when they picked up and dropped off medical reports at the hospital.

One otherworldly space I'd visited at Mass General was its historic Ether Dome amphitheater, found at the top of numerous creaking staircases. It was here, after the first public demonstration in 1846 of anesthesia's godsend for surgery, that Dr. John Collins Warren had declared, "Gentlemen, this is no humbug!" Somewhere deep down in the hospital, I'd heard, lay the morgue, once frequently referred to as "Allen Street," the former name of the street outside its door. As Lewis Thomas observed in a poem by that name, no one ever dies at Mass General; instead, "He simply sighs, rolls up his eyes, and goes to Allen Street." To this day a related euphemism is still occasionally uttered at the hospital, where "to Allen Street" someone is to pronounce them quite dead.

From an early age, and no doubt influenced by my family's business, I imagined I'd become a doctor. A heart doctor. How ever did the heart beat over and over without being plugged into a wall socket? But after I began college, the doctor idea swiftly vanished, my attention caught by the nascent field of molecular genetics that was stirring all around me. You could tell something big was afoot. It had been sown over a century earlier by the Augustinian monk Gregor Mendel, who, after years of patiently crossbreeding varieties of pea plants in his monastery garden in Moravia, proposed in 1865 that pairs of infinitesimally small entities hidden in his plants accounted for how one generation passed on its traits—wrinkly or smooth skin, green or yellow hue—with discernable probability to pea offspring.

Genetic linkage. Restriction enzymes. Thanks to a vanguard of molecular scientists at Rochester, I knew my way around these molecular tools. They were among the chief implements enabling scientists to reach into DNA, the molecule that genes are made of, although thus far relatively few genes had been isolated from organisms. Genetic linkage—a beautiful scientific truth deduced in the laboratory of Thomas Hunt Morgan at Columbia University in the early twentieth century—was the cornerstone of Mendelian genetics. *If any two segments of DNA continue to be inherited together through successive generations, it implies that they lie physically close together in the genome.* This observation was helping scientists to map the positions of genes on chromosomes, the DNA threads along which genes lay like occasional inches on a yardstick—mostly the genes of small organisms. At Rochester, I had applied endless rounds of linkage analysis to track the inheritance of certain genes in bacteria. Countless generations were needed. This entailed feeding thousands of the little critters, making them happy and getting them to mate, plating and replating their multiplying colonies, each generation separated by some twenty minutes.

Restriction enzymes, another indispensable tool, instead applied to the newer wave of genetics—genetic engineering. They amount to tiny catalyzing chemicals that act like scissors, cleaving DNA. Those used by scientists mostly come from bacteria. Should a virus invade a bacterium, the bacterium's restriction enzymes cut the virus's DNA at specific sites, thus "restricting" its ability to replicate and take over. Beneficial bacteria in our intestine, for example, rally these enzymes to disable threatening pathogens. When scientists first isolated them from bacteria in the

1970s, the driving idea was that by using restriction enzymes to cut and splice DNA, bacteria could be employed for yet another purpose: Scientists could insert human genes *into* bacteria, and as bacteria rapidly replicated, they would create scads more copies of human genes, and thus significant quantities of therapeutic human proteins such as insulin for diabetes and growth hormone for growth disorders. But as scientists got more versed in snipping long strands of DNA into smaller pieces, it also seemed a fine idea to simply study genes belonging to the human genome.

At Rochester, drawn to genetic engineering, I'd learned how to use restriction enzymes to cut—or digest—DNA; how to glue pieces of DNA together with other types of enzymes; and how to measure varying lengths of DNA. Altogether, I'd gained infinite respect for the cold logic of genetics. Its measurements had the ability to replace guesswork in science with a good degree of predictability.

The Monday following the Holiday Inn gig, I dropped by Mass General's personnel office to inquire about the opening in James Gusella's lab. Directed to the Genetics Unit, I found myself face to face with Gusella himself. Tall and bear-framed, he didn't appear too much older than my own twenty-one years, but his short brown hair, black thick-framed glasses, new jeans, plaid cotton shirt, and Serious Scientist demeanor were so far removed from my frizzy black shoulder-length mop, ragged mustache, disintegrating jeans, and Grateful Dead T-shirts—the last two items replaced that day by a suit—that I immediately decided we probably wouldn't click and I wouldn't get hired.

Gusella was from Ottawa, I learned, and had gotten his Ph.D. in biology from the Massachusetts Institute of Technology (MIT) the previous June. With open enthusiasm, he described the experiment he was directing, which, if the funding fully materialized, would require several technicians. Grand in design, it consisted of an ingenious shortcut to finding and identifying genes—or more precisely defects in genes—connected to human disease. Since the technique for locating bacterial genes I'd learned at Rochester couldn't be applied to humans—thousands of humans can't be easily mated in petri dishes and new generations don't come along every twenty minutes—a shortcut method for plunging directly after a human gene made exquisite sense to me. Clearly, the project's future ramifications for human genetics and medicine were immense, particularly since researchers were only just realizing how many human disorders arose from faulty genes.

Traditionally, the medical community attempted to understand a disease's origins through its symptoms and through its degradation of tissue and organs. But in truth, both are as distant from an inherited disease's origination point in the genome as a splattered raindrop is from a black cloud. A more recent, more exacting approach, but one with limitations, was to isolate a protein associated with a specific disease and work backwards from its structure to identify the gene that gave rise to it. Disease-related proteins were identified by inspecting the fluid or tissue affected by the pathology and noting those proteins that are abnormal in their amount and/or nature. If a protein's corresponding gene indeed contained a flaw, it just might be the disease's initiator. The mutated gene for sickle-cell anemia, for instance, had been arrived at this way, by backtracking from its hemoglobin-associated protein.

Geneticists were held back, however, because proteins related to the vast majority of genetic diseases hadn't yet been identified, so their genes were impossible to isolate. Moreover, a disease can skew the regulation of all sorts of proteins as secondary effects. As Jim Gusella today notes, "One particular protein difference doesn't promise to get you to the gene that's primarily responsible for the first, all-important change that goes wrong." As for microscopes, even though their power was rapidly improving, they were still too weak to home in on genes. Microscopes did, however, help spot blatant problems related to chromosomes, such as when an extra copy of all or parts of chromosome 21 results in Down syndrome.

"It's an incredibly powerful, revolutionary concept," exclaimed Gusella during my job interview—"to find a disease gene without any prior information about either its protein or its location in the human genome." But the concept hadn't yet been put to the test, and that was the crux of the project—to attempt to identify the genetic defect behind one particularly cruel inherited disease: Huntington's chorea, today known as Huntington's disease. Did I know much about Huntington's? I admitted I didn't, although I was aware that one of my early heroes, folk musician Woody Guthrie, had died of it.

Huntington's, untreatable and fatal, undermined brain cells associated with motor control, causing a gradually worsening movement disorder, Gusella explained. An involuntary restlessness and jerking took over an individual's arms, legs, head, and torso. Thus the flailing "chorea," or dance, described in 1872 by physician George Huntington. From onset to death, the illness often wore on for ten or more years, a victim remaining all too aware of his or her condition. Unlike Alzheimer's disease, dementia in Huntington's didn't arrive to cut the mind loose until the end stages. The only good thing about Huntington's, Gusella pointed out, was that it was fairly rare, despite the fact that if a person carried its always-causative autosomal-dominant mutation, each of his or her offspring had a 50 percent chance of inheriting it and also falling prey. Since the disorder usually didn't make itself known until between the ages of thirty and fifty, a person had already had children and had passed on the flaw unknowingly.

The attempt to unearth Huntington's deficient gene, I learned from Gusella, was in large part because of the early groundwork achieved by Woody Guthrie's widow, Marjorie. Founding the Committee to Combat Huntington's Disease in 1967, she had worked tirelessly to draw attention to the disease, help those in its path, and gain federal assistance. Among those aiding her efforts was the family of Leonore Wexler, an exceptionally gifted woman whose crippling encounter with Huntington's had spurred her family into action. Well before Leonore's death in 1978, her ex-husband Milton Wexler had established the Hereditary Disease Foundation with the goal of supporting the scientific community's invention of a useful treatment, even a highly experimental method, for halting Huntington's.

In 1979, MIT molecular biologist David Housman, known for his brilliant strokes in cancer research, had articulated a method for shooting straight into DNA after a disease gene. The theoretical approach, Gusella told me, had been brewing in various laboratories since the mid-1970s. It involved finding markers in the genome that could divulge on which chromosome a disease-inciting gene sat. Across the human species, the human genome is largely similar. But variations, it was being noticed, sometimes crop up in its 3 billion bases. A DNA variation might be as small as a one-base deletion, addition, or substitution; or as long as a stretch of thousands of extra repeating bases. Variations might sit right in a gene. (This is why each human gene comes in so many different versions, or *variants*, since a single variation makes for a different version.) Or they also could occur in nongenetic DNA, the vast stretches along chromosomes that don't code for proteins. (See figure 1.1.)

Some DNA variations were thought to be harmless. They didn't provoke disease. Others, however, did—certain mutations and polymorphisms. The incredibly bold new notion, Gusella explained, was to use a benign variation to track the harmful type. You couldn't easily single out a disease mutation; so little of the human genome had been read that wrong sequences didn't stand out from right ones.

But you could, perhaps, single out *random* DNA variations by comparing genomes from the population. And if, through genetic analysis, you found a random variation that nearly always turned up in generations of family members who have the disease, yet was mostly absent in other family members who had escaped the disease, you could infer that the found variation sat right near the disease mutation in the genome's fixed span of bases.

The linkage work I'd been a slave to at Rochester came flooding back to me: *If any two segments of DNA—say, two genes, or, as in the model Gusella was describing, a DNA variation and a gene mutation—continue to be inherited together through successive generations, it implies that they lie physically close together in the genome*. They sat *so* close together—on the very same strand of DNA—that the two rarely got separated at meiosis, the point at which sperm or egg cells divide and a person's two copies of chromosomes exchange and recombine genetic material, this recombination passed down to offspring.

Housman's suggestion of a gene hunt, which he shared with the Wexlers as well as Joseph Martin, Mass General's then chief of Neurology Service, sparked a bonfire of like-mindedness. Martin was tremendously keen on the proposal, as was Nancy Wexler, one of Leonore's daughters and soon-to-be director of the Hereditary Disease Foundation. The project became reality once Mass General applied for and received funds from the National Institutes of Health that were available due to a congressional incentive to support advancements against Huntington's. Housman recommended to Martin that his standout student—Jim Gusella—serve as the venture's principal investigator.

Nancy Wexler, meanwhile, was aware of a large family in Venezuela with a history of Huntington's. She volunteered to lead a medical

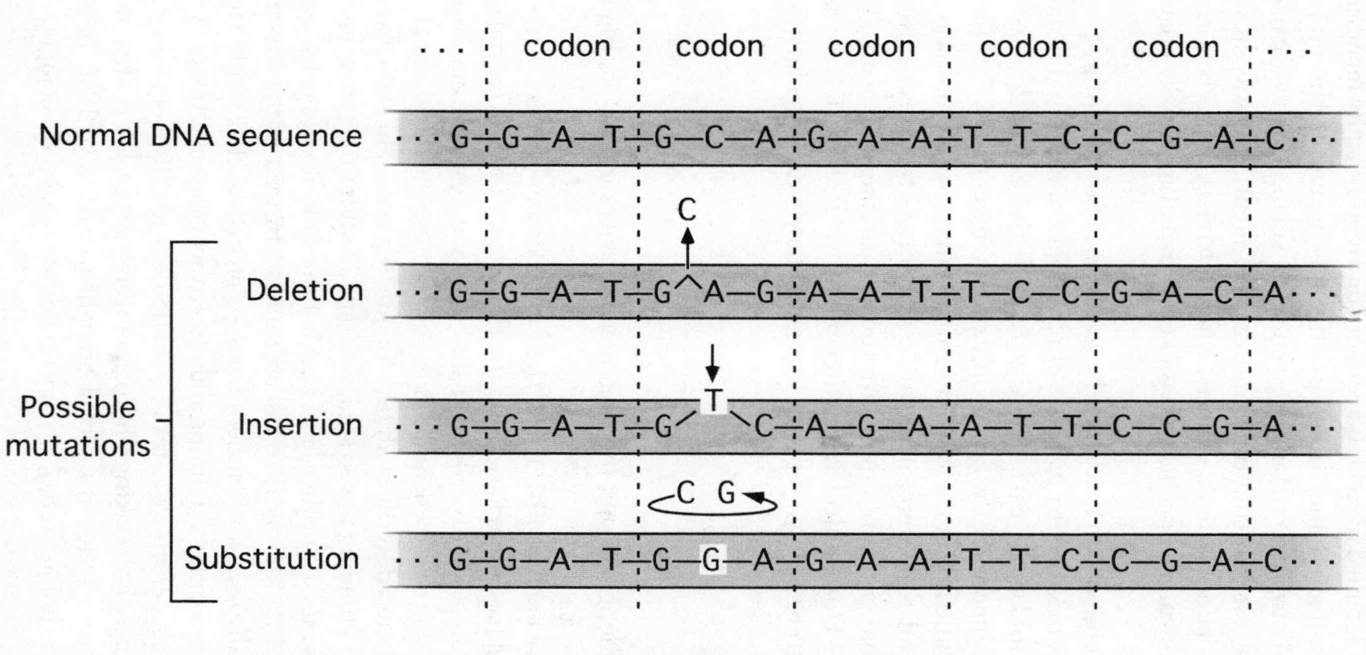

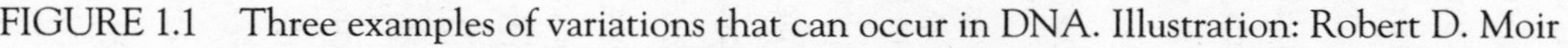
FIGURE 1.1 Three examples of variations that can occur in DNA. Illustration: Robert D. Moir

team to South America to gather blood samples from its members. Extensive DNA culled from blood cells of Huntington family members—both those affected and those not affected—would be the project's most crucial raw material, and the Venezuelan clan, which were manyfold the largest Huntington's kindred ever identified, was an ideal source. In a poor Catholic country, where a woman might bear ten or even twenty children, *el mal* (the sickness) had all too easily engulfed a multitude.

Gusella ended our interview by giving me fair warning. Simply linking Huntington's to its home chromosome, never mind isolating its gene, could take years. There were even those who doubted the whole concept would fly. So far only *one* anonymous DNA variation had been pulled from the human genome, and since no one was sure how plentiful they were, it might be unrealistic to expect to find a variation that got coinherited with the disease time and again. "But I have faith," Gusella declared matter-of-factly. "People's DNA varies. There has to be a polymorphism located somewhere in the vicinity of Huntington's gene."

Despite my long hair and the counterculture vibes it gave off, I landed the position.

As I discovered my first day on the job—October 1, 1980—Gusella's two-roomed base of operations on the third floor of Mass General's Research Building was exceedingly bare-bones and dusty. At our disposal were three short lab benches, a hood for venting chemical fumes, and two glass-enclosed sterile chambers for growing cells. That was about it. The project's funding was so slim, we could ill afford the pricey biotech merchandise used in other labs. Handed the task of physically setting up the lab, I had to resort to resourcefulness. Instead of the expensive trays normally used for Southern blots—a procedure that would enable us to both recognize variations and discern whether one traveled with Huntington's defect—we initially made do with the bottoms of old hamster cages. For making photographic images of DNA that would ensure our tests were on target, I brought in a vintage 1966 Polaroid camera from home. Today, my unit relies on a $20,000 computerized system to document DNA gels. But back then a $25 Polaroid and wrinkled red acetate swiped from my band's light-show equipment did the job.

Gusella brought aboard two more assistants. Mary Anne Anderson, the first to arrive, had training in tissue culture, the art of feeding, growing, and nurturing human cells. Her proclivity in this regard was an absolute necessity, since our tracking of Huntington's mutation would require the establishment of hundreds of cell lines—each a sample of a person's blood cells that can be grown indefinitely. This is done by introducing a cancer virus, usually Epstein-Barr. Immortalized cells would ensure unlimited supplies of DNA from the two Huntington families our experiment would draw from—the large Venezuelan clan as well as a much smaller family from Iowa, whose blood samples had been collected by P. Michael Conneally, a geneticist at Indiana University School of Medicine.

Next to arrive was Paul Watkins, a giraffe-tall student of David Housman at MIT. His ongoing master's thesis—crafting techniques to identify variations in DNA specifically on chromosome 21—also made him a perfect addition to the team. While he and I worked away at jerry-building various nuts and bolts, Gusella wrestled with the project's technical design. If we were to make progress along the genome's interminable highway, our DNA tests had to be run on a much larger scale than was the norm in those days. Gusella and Housman—"two higher life forms," as Watkins remembers them—regularly got together to hammer out the intricacies.

As Gusella had pointed out, the most remarkable thing about the Huntington's search was that we were sailing into DNA's sea of 3 billion bases without a clue about the Huntington gene's location or even a compass. The gist of the whole experiment was that we'd analyze random pieces of DNA from the general population for variations, and, as long as we found them, we'd track the inheritance of each of these markers in the Huntington families' DNA. But quite possibly all the doubting Thomases might be right. Even if we dug up variations, we might never find one particular one that sat physically close to Huntington's mutation, hence could lead us there.

Working up to speed and counseled every step of the way by Gusella, Paul Watkins and I began hunting down variations. We were employing the innovative technique that had convinced Housman and Gusella that variations were findable. This brand new trick of the trade relied heavily on the scissory action of restriction enzymes, for it was *where* these minuscule motes of chemicals chose to cut DNA that would reveal variations. Each type of restriction enzyme recognizes and cleaves DNA at specific sites, which results in DNA fragments of a predictable length. But if a variation exists in a person's DNA, longer- or shorter-length DNA fragments may result, and these irregular fragments can signal a variation.

The multistep procedure of setting restriction enzymes loose over DNA, then recognizing the handiwork they've left behind, constitutes what's called a Southern blot—a 1975 invention of Edwin M. Southern. Blotting entails cutting up, say, six random people's DNA with various restriction enzymes, inserting the resulting fragments into a gel, then zapping the gel with an electrical charge, which separates DNA pieces according to length. Once the fragments are transferred onto solid filters, you're left with a semipermanent framing of a spectrum of differing-length fragments. You then randomly choose a piece of DNA from a genome library—an anonymous person's genome that's been cut up into thousands of pieces. You turn this little piece into a probe by making it radioactive, then run it over the six people's cut-up DNA, allowing the probe—which could come from virtually anywhere in the genome—to hybridize, or bond, with its like pieces. By doing this, in essence you are tagging the same short sequence of genome from all six people. When you put these tagged pieces on film (an autoradiograph), each one appears as a black band—your visual cue. If the lengths of the bands differ in any of the six people, you just may have found a variation that occurs in the general population.

More rounds of Southern blots, this time applied to Huntington family DNA, would ultimately provide a reading of each person's two alleles, or gene copies and tell us whether a specific variation in an allele occurs more frequently in the DNA of those who have the disease and whether it is passed down from affected parents to affected children.

Many of our early attempts at Southern blots failed, blowing days of work and wasting valuable reagents, especially DNA. Flimsy filters collapsed. Or valuable pellets of DNA slipped off the end of pipettes, lost to the floor. Anderson would have to grow up the DNA all over again.

These early failures were immensely frustrating, since presumably hundreds of variations would have to be identified before we locked onto one that just happened to sit next to Huntington's faulty gene. The immediate plan was to dredge up several, if possible, and test them against the DNAs of the Iowan Huntington family. About to arrive in the lab was the first sizable shipment of the bloods of the Venezuelans, which Nancy Wexler and teammates were collecting, and we'd next run any landed variations against that kinship's DNAs.

Gusella, despite a full load of administrative and technical responsibilities, often pitched in at the bench, offering sound advice. My apprehension over working for someone so seemingly different from myself quickly vanished. He had an uncanny ability for grasping complex ideas and for recognizing a compelling hypothesis in abstract data. Just as impressive was his nimble versatility at conceptualizing on most any subject, from genomics to politics to baseball. A hunger for logic had originally attracted him to science, and it was this penchant for the rational and the reasonable that, right from the start, carried us cleanly over many a technical obstacle. His entire philosophy of life, he'll tell you today, is based on Walter Brooks's "Freddy the Pig" children's books, filled as they are with fanciful, logical practicalities. One of his favorite Freddy-isms hails from the time another farm animal comes to visit Freddy and inquires why Freddy's windows are so dirty. Freddy's response—sounding very much like the Gusella I've come to know—is that if his windows were clean he'd know who was walking up to his house, and it would ruin the wonderful surprise of a knock at the door.

Although he was a fish in water in the lab, Gusella could be shy and even uncomfortable in social gatherings. If he was in Heidelberg for a genetics meeting, rather than join the evening cocktail circuit, for instance, he preferred to get in a car and drive around, stopping for pizza or aiming for the McDonald's in Heidelberg's famous pink building, but avoiding fancy restaurants at all cost.

By early spring 1981, Watkins and I were beside ourselves. Among the first five probes we'd used from an anonymous human genome, four—

G3, G6, G8, and G9—had yielded one-base variations in the DNA of the general population. In fact, G8, which was many thousand bases long, had yielded three variations! Since we had caught sight of so many variations this soon, Gusella must be right—the genome must be loaded with them, making it all the more believable that eventually we'd stumble onto one that coinherited with Huntington's flaw.

Doing science by day while making music at night left time for little else. Often I would leave the lab at dusk, hook up with The Nunz—a punkabilly rock band I'd joined that played a blend of new wave and bluegrass—return home in the wee hours to catch three winks, just make the 7:20 A.M. Providence-to-Boston bus the next morning, and crawl into the lab before the first pot of coffee had receded to a burned puddle. I often regaled my labmates with horror stories from the night before—most memorably, the night that the band's music-timed gunpowder bombs nearly incinerated a roomful of dancers. Whenever free moments allowed, I took refuge in composing music. My songwriting partners Colin Wheeler and Charlie Lavalle added the lyrics, and we'd send off our creations to local radio stations like Providence's WBRU and Boston's WBCN.

By early 1982, Mary Anne Anderson, frequently helped out by Gusella, was extracting enough DNA from the cell lines of the Iowans and Venezuelans to allow us to start testing our found DNA variation markers against the DNAs of these two kinships. Although we felt more assured that we would bump into an informative variation for Huntington's, so far we'd captured only six variations, so "someday" still might be years off. As long as our funding didn't run out, this wasn't of any great concern. No other teams were in pursuit of the gene at that point. And for each of us it was a magical time in many ways. Beneath the drone of benchwork lay a constant anticipation and the knowledge that if we really did succeed in hooking Huntington's mutant gene, our example would catapult medical research into a new era.

"It became assembly-line science—setting up the day's experiments, pipetting up ten solutions, labeling hundreds of test tubes, and subcloning DNA," recalls Kathleen Ottina, a biochemist who joined the team at the start of '82. "But it was incredibly exciting and I loved the hands-on aspect, something that today's automated lab has lost. There was a romanticism attached to doing it that way. It was like Marie

Curie boiling down pitchblende in her backyard laboratory to purify radium."

Due to another expedition by Nancy Wexler's group to South America, hundreds more of the Venezuelans' blood samples began crowding the lab in the spring of 1982. The lab started to resemble an ocean-liner that pushes its engines to max knots upon reaching open sea. New technicians were hired to help transform cell lines, cut up DNA, and process the endless Southern blots that were helping us to gradually reveal variations in normal DNA and check them for linkage in disease-ridden DNA. The processing of Southern blots was my bailiwick. Because of my late nights stomping on some stage, I was often on the verge of double vision by midafternoon, and to stay alert, I'd fantasize that I was a contestant in a molecular-biology Olympics. To beat out the invisible competition, I had to do one blot after another as quickly yet as perfectly as possible.

I began wondering if my lack of sleep wasn't making me sloppy at the bench, because toward the end of 1982, something very curious was happening. Two of the variations we'd found in the little G8 piece of genome showed a hint of linkage to the disease in the Iowans! This was the outcome we had pinned our hopes to, yet it seemed unbelievable since the odds this early in the game practically defied plucking out a variation that showed linkage. As Gusella recently had promoted me to senior technician, I felt doubly accountable for any blunders. Gusella, meanwhile, alternated between curiosity over the results and suspicion that one of us had indeed botched an experiment.

While eyebrow-raising, the G8 probe's linkage to Huntington's wasn't spectacular. Its lod score, a statistical indicator of whether linkage between two points on DNA is significant, was being computed by Peggy Wallace, a grad student in Mike Conneally's Indiana University lab. Quite literally, it's a log (logarithm) of the odds. A lod score of 3.0 and over was deemed very significant, yet G8 in the Iowans had reached only 1.8.

Gels, blots, hybridizations, autoradiograms—round after round continued as we pressed on to analyze other variations we had alighted on. We realized that we still might have thousands of rounds left to go—cleave, zap, blot, probe—before we turned up a definitive marker. Cell lines were accumulating in the freezers; the counters were perpetually covered with gels, blotting trays, and dozens of mixtures, not to mention the countless little white stalagmites that grew up when the salty solutions used for Southern blots spilled and dried. Today, you can special order just about every procedure and reagent—even DNA from families with a specific disease—but back then most everything had to be painstakingly hand-prepared.

Not that we were without diversions. Inseparable by this point, the team nearly always lunched together, letting off steam in Mass General's dingy basement cafeteria, or the Mass Eye and Ear where the food was better, or our favorite deli—the Metro on Cambridge Street. Often after lunch, Gusella and I—Paul Watkins by then having finished his master's and taken a biotech job in the suburbs—walked over to the Boston Garden's video arcade and played Pacman, Zaxxon, or table hockey for as long as time allowed. Back at the bench, midafternoon trivia jousts helped relieve repetitive-motion syndrome.

When data from a small subset of Venezuelans were added, G8's score rose to 2.2, providing nearly 300 to 1 odds that we had locked onto the Huntington's gene. Eyebrows arched even higher. Were we kidding ourselves? One of our first used probes, which had been blindly pulled from the genome's 3 billion bases, had yielded a variation marker that was delivering? Gusella remained cautious, downplaying our linkage data at a Huntington's session at the World Federation of Neurology meeting in Chicago in April of 1983. Nothing but a lod score of over 3.0 would convince him.

To reach statistically firmer ground, we had but one option: to test G8 against more Venezuelans. Cleave, zap, blot, probe. But various delays began slowing us down. We ran into endless Southern blot snafus; and all too frequently radioactive contamination left over from the hybridization process obscured our autorads' discrete black bands that denote an individual's two alleles of a gene. Gusella and I would be in the darkroom, on the verge of developing key evidence, only to hold the film up to the safe light and see—instead of bands—"cat vomit," large black clouds that obscured the image.

Gusella and I decided to grow up fresh DNA from roughly 150 of the Venezuelans' cell lines and do all the cutting, blotting, and probing ourselves. The two of us could exert controls over the procedures that were getting lost in the group effort. In mid-July, over the course of several days, we plowed straight through all the necessary steps. On a Wednesday we got together to process the autorads, trying to be patient and not pull them out of the developer solution too soon. As we held them up to the safe light, the moment of truth arrived. Hurray, no cat vomit! The bands were clearly decipherable. Later that day, while I read off their data, Gusella, poker-faced and murmuring an occasional "okay, okay," pencilled in the genotypes—who had the G8 variation, who didn't—onto the Venezuelans' family tree. As if not to jinx us, he didn't share his observations, but immediately FedExed the data to Peggy Wallace in Indiana for computer analysis.

At home that night, in the north Providence apartment I shared with my fiancée Janet, I sat cross-legged on the living room rug watching *Love Boat* with the volume lowered while listening intently as WBRU counted down the local top ten singles. "New Plan," a song I'd composed with Colin Wheeler, had climbed to number five on the list, lifting me cloud-high. Spread out before me was the Venezuelans' family tree, and as I studied it I could see for myself that every Venezuelan with the disease had inherited the G8 variation from their affected parents; those without the disease had not. Whoa! A science-to-music grand sweep? Could it really be?

It would take several days for Conneally's lab to assess the data's lod score. But Gusella, having already scanned the autorads by eye, recalls, "I didn't have to wait for the computer calculations to know we had linkage." He, his wife, and toddler son headed north to Ottawa for a brief visit with relatives. "The drive was incredible," he recalls. "As we passed through the mountains of New Hampshire, I had a hard time staying on the road. You could just see that the finding was going to cause human genetics to explode." Definitive word arrived from Conneally: the additional data from the 150 Venezuelans had raised the lod score to over 6.0! That placed the odds of having linkage at 1 million to 1.

The G8 probe, it was soon determined, came from the short arm of chromosome 4. Bang! The Huntington gene must also reside on chromosome 4. It might take many more months, even years, to know which of the hundred or so genes that lay in the same region were defective in Huntington patients and identify the exact DNA error. Yet the chromosome's capture alone was fantastically historic. For the first time, the approximate location of a human disease gene had been found with

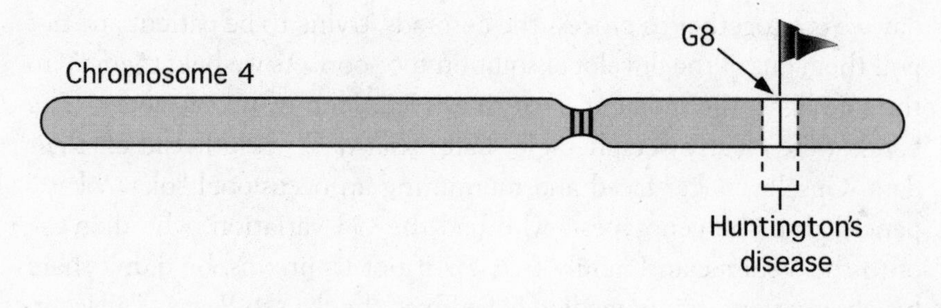

FIGURE 1.2 The site of marker G8's linkage to the Huntington's gene. Illustration: Robert D. Moir

absolutely no prior clues about its genomic address. Our approach had proved to be sound. The biology of human disease had been cracked wide open, rendering the human genome an examinable part of the anatomy and potentially laying bare the genetic roots of thousands of other inherited troubles.

The British journal *Nature* carried Gusella's and Joe Martin's formal report of the breakthrough in its November 17, 1983, issue. Around the world headlines paraded the sensational news. All but overnight, Gusella went from being an unknown Ottawan to a world-famous geneticist touted far and wide as "Lucky Jim"—although to this day Gusella maintains that if latching onto G8 so early in the project was luck, it was "well-reasoned luck, because we prepared for it."

The instruction book for human life, crammed with zillions of secrets, lay wide open for exploration. And I was hooked. The ability to directly expose the genetic basis for a disease made what every medical researcher dreams about all the more real—parlaying deep-seated information about an inherited disease into an effective treatment or a cure. Having been captive to the Huntington's capture, I saw my beliefs widen and stretch. Success left me hungry, but for what I wasn't entirely sure.

During the spring of 1983, just when Jim Gusella and those of us in his lab were striving to get a clearer sense of G8's linkage, a veteran pathologist at the University of California, San Diego, was making unprecedented headway into the molecular fracas of a totally different fatal neurologic disease. It's uncanny, in some respects, how close together in time these two separate advances occurred. Not far into the future they'd intersect, the second discovery made all the more promising because of the first.

Julia Tatro was fifteen when she first laid eyes on the blond, blue-eyed lifeguard at Campello swimming pool in Brockton, Massachusetts. His name was Johnny Noonan, and before long she was crazy about him and turned his head—this shapely Tatro twin whose five-foot-three-inch frame spun with energy and whose smile and laugh carved a deep dimple on her left cheek. Both were students at Brockton High, and so became high school sweethearts. Johnny was a New England diving champion, and when they went to dances together, he was paid to do diving demonstrations during intermissions, making Julia all the prouder of "Champ," as his friends called him. After high school, they married. Johnny worked as a telephone lineman, switched to insurance, then eventually joined the Stoughton fire department where he rose to lieutenant. They had their first baby, then a second. Of true Irish-Catholic stock, they then had a third, a fourth, and a fifth. Julia loved bearing children, watching their idiosyncrasies emerge, and hearing their chatter. As their family expanded, filling their barn-red house with voices, her children remember her singing, "Oh Johnny, oh Johnny, how you can love!"

— *two* —

The Core of the Matter

We dance round in a ring and suppose,
But the Secret sits in the middle and knows.
—Robert Frost, "The Secret Sits"

When George Glenner glanced down the barrel of a polarizing microscope one June morning, his swift assessment was all-encompassing, his approval immediate. The vivid field of apple-green below his eye was as uniformly bright as he'd hoped it would be. The brighter the better. The night before, Glenner's technician Cai'ne Wong had added a stain called Congo red to a small glob of grayish debris extracted from blood vessels of brains ravaged by Alzheimer's disease. The resulting apple-green indicated that after months of improving their methods the two scientists had succeeded in isolating an enriched sample of *amyloid.* Amyloid, which Glenner viewed as "one hell of a nasty substance," is the gummy protein that abnormally, relentlessly accumulates in the brains of Alzheimer victims. That the older and younger scientists had wrested a bit of it free from other brain material was definite progress. But as Glenner was keenly aware, it was a far cry from their bigger goal, which was to decipher the protein's component parts, its amino acids.

At that time—the spring of 1983—I hadn't yet heard of George Glenner. In fact, I'd only recently become acquainted with amyloid. This introduction occurred when Jim Gusella and I attended a lecture by an up-and-coming Alzheimer researcher—Dennis Selkoe, then working at McLean Hospital outside Boston. As a person succumbs to Alzheimer-related memory loss, disorientation, and other symptoms, Selkoe described, millions of amyloid's microscopic plaques invariably are forming and lodging between brain cells like strewn boulders in prime regions of the cerebral cortex. Amyloid also is collecting in the brain's blood vessels, clinging to their walls like barnacles. Selkoe spoke about his recent discovery of amyloid's terrific insolubility; how, unlike many proteins, it was incredibly hard to break down in solution, which thwarted attempts to obtain its chemical composition. Driving home his point, he showed a slide of Alexander the Great cutting the Gordian knot. "In Greek mythology we read about this terrible twisted knot that no one could unravel," said Selkoe, "and Alexander the Great came along and in one fell swoop cut it with his sword, and no one had ever thought of doing it that way." It was the same with amyloid, Selkoe maintained. Because it couldn't be easily untied, it had to be gone at decisively if anyone hoped to read its sequence and learn its makeup.

Selkoe's talk left me with two strong impressions about brain amyloid. At the biological level there was no mistaking this fiercely insoluble protein as Alzheimer's most flagrant intruder. And quite possibly something about amyloid might even cause Alzheimer's, although why it ran amok in the brain had baffled researchers for a very long time.

As I would learn in the years ahead, George Glenner firmly believed that amyloid played a central role in Alzheimer's extensive death of brain cells. His hunch was but a hunch. It could be that amyloid deposits, or plaques, were nothing but "tombstones," as some called them, refuse deposited by the disease due to a pathological event unrelated to amyloid. The best way of getting some answers, decided Glenner, was to cut to the chase—to dig the protein out of the brain and analyze its chemical makeup. Not surprisingly, other scientists similarly envisioned the huge rewards for medicine. They too were trying to separate out Alzheimer's insidious protein from postmortem brain samples, then sequence it. The task, however, was hardly an easy one.

Glenner considered amyloid one of the great all-time medical mysteries, and for decades he had doggedly, obsessively tried to fathom everything he could about it. The amyloid he had devoted most of his career to, however, wasn't the Alzheimer type found in the brain. Instead, he had inspected numerous other types that occur in other organs throughout the body, almost always with disease consequences.

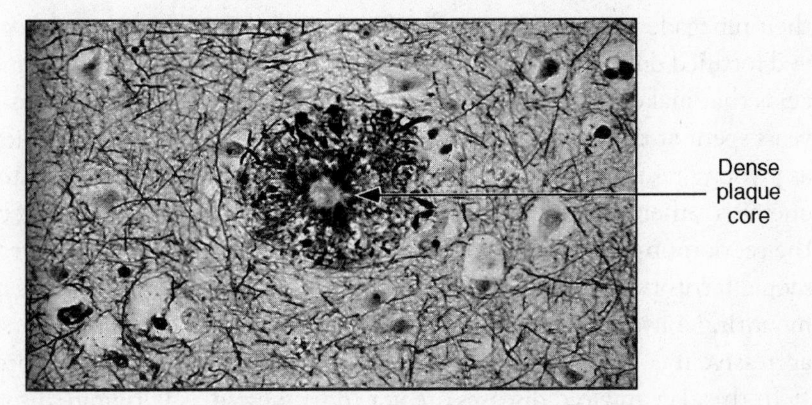

FIGURE 2.1 Photomicrograph of amyloid plaque in the cerebral cortex.

Born in Brooklyn, New York, on September 17, 1927, and raised on the Main Line outside Philadelphia, Glenner had pursued his boyhood fascination for medicine by earning an M.D. from Johns Hopkins University School of Medicine in 1953. He intended to become a surgeon, but during his internship at Mount Sinai Hospital in New York, he began to have serious doubts about his chosen field. When he was eight years old, meningitis had left him with only partial hearing in one ear, and though he wore a double hearing aid and employed a special stethoscope designed for the hearing impaired, his deafness set him apart in the clinical setting. "Say it again," he repeated time and again. Aside from his disability, he was put off by surgeons' high fees. It was hard for him to imagine charging patients so much money to make them well.

Like many others entering medicine in the postwar years when a prosperity-driven impetus toward healthcare abounded, and when new vaccines and antibodies were raising optimism that many dire illnesses could be addressed, Glenner felt the lure of looking deeper into the cellular world for answers about disease. That he received his M.D. in 1953 says something about the building expectations all around him. That year James Watson and Francis Crick—an American biologist and British biophysicist—set the biology world on its ear when, with an eye toward X-ray pictures of DNA taken by crystallographer Rosalind Franklin, they discovered DNA's double-stranded, coiled structure.

Pathology, which Glenner switched to in the mid-1950s and trained in, gave him license to try to get to the bottom of deadly disorders and their renegade proteins. Due to the chemistry he'd majored in at college and fortified during med school, he was no stranger to the twenty amino acids that make up all human proteins. For the better part of twenty-six years spent at the National Institutes of Health (NIH) in Bethesda, first as a young research assistant in the Laboratory of Pathology and Histochemistry, then as chief of Histochemistry, then finally as NIH's chief of the section on Molecular Pathology, he delved ever deeper into microscopic territory. He fixated especially on the chemical characteristics of mounting amyloid proteins inside various organs, an accumulation so aggressive it could inflate the liver, for instance, to twice its normal size.

To this day amyloid diseases are apt to be viewed as fairly rare disorders. Most in fact are. Yet if the entire spectrum is taken into account, including the many cases associated with type II diabetes and Alzheimer's disease, this pathology ranks as anything but rare. In some instances, amyloid's profuse lesions turn up as a secondary consequence of inflammation—for instance, in lungs plagued by cystic fibrosis or in joint cartilage inflamed by rheumatoid arthritis. Even leprosy and malaria can incite the excessive formation of amyloid. In other cases, amyloid itself constitutes the primary disease-causing material, as seen in cases of cardiac amyloidosis wherein amyloid thickens the walls of the heart's chambers and befouls vessels leading in and out of the heart, usually with fatal consequences. All told, no organ or tissue type appears immune to amyloid's exigencies. It can build up in the kidney, liver, pancreas, heart, spleen, lungs, bladder, bowel, penis, and even in the eye, not to mention the brain. Skin, including facial skin, can become scarred with its lesions. Vocal cords, hence speech, can be altered by it.

Glenner's attempt to tease out amyloid's properties particularly paid off in 1968. Many different types of amyloid proteins exist, and collaborating with David Eanes, an NIH colleague, Glenner made the discovery that each and every such protein folds into one of the most common three-dimensional protein configurations found in Nature—a beta-pleated sheet, a pattern not unlike that of a pleated kilt. Alan Cohen, an amyloid authority at Boston University School of Medicine, soon reported the same finding. In humans, when a normal, soluble protein with a helical "alpha" confirmation converts into this zigzag "beta" confirmation, it becomes wickedly insoluble, which explained why once amyloid amassed, the body could not break it down and get rid of it. A beta-pleated structure might be abundant in Nature, but too much of it in the body obviously could be a liability. Since amyloid proteins shared this structural commonality, their associated diseases were seen as all the more of a syndrome.

Over a century earlier the preeminent German pathologist Rudolf Virchow wrote, "Only when we have discovered the means of isolating the amyloid substance, shall we be able to come to any definite conclusion with regard to its nature." Due to pioneering work by Glenner and others—notably Earl Benditt at Washington University, Seattle, Edward Franklin at New York University, and Alan Cohen at Boston University—a range of amyloid proteins associated with different diseased organs would be scoped out. Today, as many as twenty stand identified. Although amyloid proteins differ in amino-acid sequence from disease to disease, each follows the same ruinous course in human organs: as amyloid lesions multiply, they crowd out or compromise surrounding cells to the point where normal cell functioning becomes degraded or destroyed. The central question that would arise in relation to Alzheimer's disease was, Did excess amyloid in the brain do the same?

It wasn't until 1977, when he was invited to talk on amyloid at a conference on Alzheimer's disease at the National Institutes of Health in Bethesda—the first such major U.S. meeting to rally researchers against Alzheimer's—that Glenner turned his focus to the amyloid deposited in an Alzheimer brain. Here, he realized, was an amyloid disorder that had been only crudely investigated, a cousin to all the amyloid conditions in other organs. Most neurologists around Glenner, meanwhile, had their blinders on. Their area of expertise being the central nervous system, they saw Alzheimer amyloid in the narrow context of a brain-related problem. "The field needed a George Glenner who really didn't know that much about the nervous system, and didn't need to, to make the key point that this disease in the brain was very similar to amyloid accumulating in spleen or bone marrow," notes Dennis Selkoe, now a Harvard Medical School (HMS) neurologist and codirector of the Center for Neurologic Diseases at Boston's Brigham and Women's Hospital. "Sometimes it takes an outsider to show the way."

Ailing organs that were enlarged by peculiar microscopic clusters had been reported on ever since Shakespeare's day. In 1853, Virchow, having studied amyloid from various tissues, named the mysterious substance. (Two years later came Virchow's famous observation that existing cells divide and produce new cells, a theory that far outshone his scrutiny of amyloid.) He chose "amyloid," from the Latin *amylum* meaning "starch," because its deposits looked so waxy and white, and changed to a bluish color just the way vegetable starch did when treated with iodine. Evidence grew that amyloid was actually protein-derived, but Virchow's term remained in use, an accepted misnomer. As Alan Cohen points out, the simple act of naming a formerly nameless material was to serve as a "remarkable unifying force" for further study. An amyloid disease would become known as an *amyloidosis*.

The serious form of dementia described by the Bavarian neuropsychiatrist Alois Alzheimer on November 4, 1906, at a meeting of psychiatrists in Tübingen, Germany, likewise was a condition in need of a name. Alzheimer—a large, dapperly attired man who excelled at exceedingly small and delicate observations of many neurologic diseases—recounted the wayward, seemingly insane behavior of Auguste D., one of his patients, in the years prior to her death: her worsening memory, her disorientation, her non sequiturs, her occasional delirium and hallucinations, her frenzied jealousy toward her husband. Alzheimer had first examined Auguste D. in November 1901 when she was admitted at age fifty-one to the Hospital for the Mentally Ill and Epileptics, the Frankfurt asylum where Alzheimer got his first appointment as a physician. Over the next few years he had followed her downward spiral, until her death at age fifty-six. As he went on to tell his listeners, a postmortem examination of her brain had revealed flagrant peculiarities. Aided by a light microscope and a newly invented silver stain, Alzheimer had detected widespread "miliary foci," or thousands upon thousands of tiny clusters strewn across Auguste D.'s cerebral cortex. Apparently he did not recognize these deposits as amyloid. He also spied innumerable neurons, or nerve cells, whose interiors were choked by "dense bundles of fibrils," as his written report later stated. Today we call these knots of filaments, which are associated with several other brain diseases, *neurofibrillary tangles*.

Here, then, was evidence of a biological basis for a form of insanity—dementia in those days was usually regarded as a psychosis, a mind disorder categorized separately from a disorder of the anatomy. What has always struck me as curious about accounts of Alois Alzheimer's presentation is that when he concluded, his listeners had no questions. As they dispersed, were any of them pondering this proof of a physical disease underlying a depraved mind? Or that any number of patients crowded into the insane asylums of the day might be suffering from this same suspicious brain dirt? Were the implications lost on the room, or was the significance of Alzheimer's findings so clear that no questions needed asking?

While others before Alois Alzheimer had noted the rampant, strange deposits in the brains of demented individuals, Alzheimer's sighting of the tangles inside cortical neurons was new. Four years after his Tübingen presentation, Emil Kraepelin, a leading German psychiatrist, named the disorder after his younger friend; perhaps because no one before Alzheimer had studied the pathology quite so meticulously. Because of the age of Auguste D. and others with the same condition, the illness was assumed by many to be a presenile dementia, one that struck before old age. The same year—1910—that Kraepelin named the disease "Alzheimer's," Gaetano Perusini, a doctor and comrade of Alzheimer who had received and read clinical reports of the Ophelia-like waywardness that gripped Auguste D., published them in a paper. The following excerpts describe Alois Alzheimer's patient in the late days of November 1901, the year she first came under his care.

> (What is your name?) 'August.' (Last name?) 'August.' (What is your husband's name?) 'August, I think.' . . . During lunch she ate cauliflower and pork; asked what she is eating, she replies; 'spinach.' . . . When asked to recite the alphabet, she arrives at G, and when asked to continue she says 'I am not dressed for this.' . . . She acts as if she were blind, touching the other patients on their faces and when asked what she is doing, replies: 'I must put myself in order.' . . . During the exam she suddenly says: 'Just now a child called.' . . . When the doctor enters the room she tells him to stay away; other times she greets him as if he were a dear guest. 'Make yourself comfortable, until now I haven't had time.' Later she becomes excited again and screams terribly, like a small child.

Scientists soon realized that the plaques accompanying Alzheimer's were predominantly composed of amyloid. But the base substance of the disease's other classic lesion—the neurofibrillary tangles—remained a mystery. For decades, right up through the 1950s, researchers lacked the means to explore the illness's deep-seated causes to any great degree, as was true of so many neurological ills. This didn't stop debate over the strange, copious litter it heaped upon the brain. Did the plaques suspended *between* neurons or the tangles seen *inside* neurons form before neurons began dying or after? Was one lesion or the other responsible for undermining nerve cells? Because the disease was usually still pegged as a psychosis, those scrutinizing it were mainly psychiatrists, who attempted to define it by taking close note of the demented behavior it unleashed. As for neuropathologists, "They were mostly occupied with diagnosing brain tumors for the neurosurgeons," according to Robert Terry, a neuropathologist at the University of California, San Diego, and one of the visionaries behind modern-day Alzheimer research.

Beginning in the early 1960s, researchers equipped with the powerful, new electron microscope (EM) finally could observe more clues about the disease. Invented in Germany in the 1930s and employed for tissue studies two decades later, EM revealed things no one had ever seen before in the cellular world and elated Alzheimer investigators with its unparalleled detail of plaques and tangles. So memorable for Terry—then at Albert Einstein College of Medicine and a pioneer in applying EM to Alzheimer's—was the day in 1960 that his attempts to pull tangles into view had worked. For Terry—a World War II paratrooper—staring down into these abnormal heaps inside deteriorating brain cells was the same as plunging down into enemy territory.

Early EM studies principally served to confirm and better describe the ultrafine subcomponents of Alzheimer's plaques and tangles. In 1963, Michael Kidd, a neuroanatomist in England, was the first to observe that the tangles inside neurons were composed of pairs of abnormally entwined filaments. Terry and his EM team meanwhile went on to distinguish the main features of the plaques. At their core lay stuck-together amyloid fibrils. Often glued to the outer fringes were masses of supporting brain cells—glial cells. Extending into and around the plaque were the swollen, dystrophic armlike extensions of nerve cells, or neurons. These normally thin fibers, which a neuron grows for relaying messages, are called *neurites*. Because of their presence in the fray, Alzheimer's amyloid plaques got the name *neuritic plaques*. "The neuritic plaque is a real mess—a complex lesion of amyloid deposition, surrounded by degenerating neuronal arms and legs and glial processes. EM helped figure out which roots belonged to which tree," noted neuropathologist Henry Wisniewski, a valuable member of Terry's team who went on to direct the Institute for Basic Research on Staten Island. (Wisniewski, highly respected for his many contributions to our field, passed away in the fall of 1999.)

Despite their gleanings, electron microscopists alone couldn't hope to solve the puzzle of which lesion—the plaque or the tangle, or their subunits—was first on the scene and whether either type lesion actually was responsible for the disorder's slaughter of neurons. In the past, the tendency was to blame the disease on the tangles, simply by virtue of the fact that, unlike the plaques, they formed in the interior of cells that died. But there was no hard evidence that tangles *killed* neurons. Could it be that the tangle, the plaque, or perhaps both lesions were mere rubble left over from a more primary insult to the cell? Another point of interest was that some older brains—healthy ones as well as those strewn with neuritic plaques—contained prodigious amounts of "diffuse" plaques. These lack an amyloid core and surrounding swirl of damaged neuronal arms. Were these diffuse clouds precursors of neuritic plaques? Were they a sign of the disease to come?

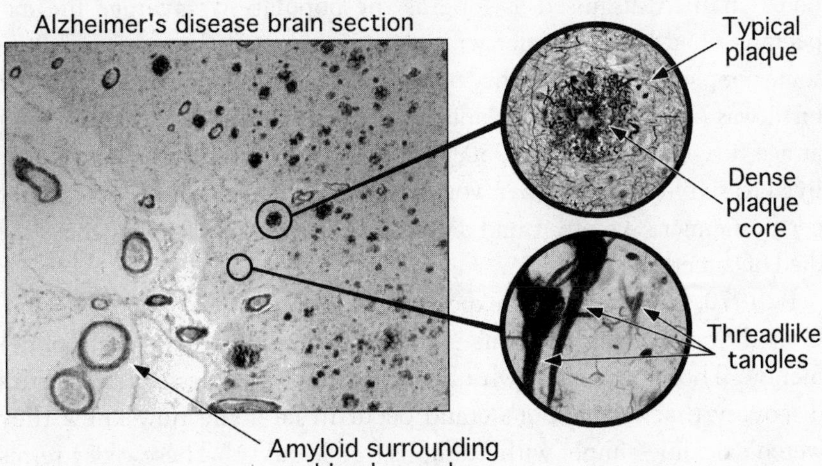

FIGURE 2.2 Photomicrograph of Alzheimer's plaques and tangles.

While electron microscopists were scratching away at these and other unknowns, a drastic change in the perception of Alzheimer's disease was under way. Ever since Alois Alzheimer had reported on the brain abnormalities associated with Auguste D.'s demise at fifty-six, the belief largely continued that the disease was a rare presenile dementia. If a person suffered the same terrible mental collapse later in life, their symptoms often were thought to stem from something quite separate. Many suspected age-induced hardening of the arteries, or arteriosclerosis, whereupon arteries compress, starving the brain of oxygen-rich blood. Others blamed simply the aging process. But these explanations began to falter in the 1950s and 1960s due to mounting autopsy-based evidence that Alzheimer-type plaques and tangles were also present in copious amounts in older brains, not only younger brains. Electron microscopy validated this insight. Three British pathologists—Bernard Tomlinson, Gary Blessed, and Martin Roth—are credited for finally unveiling Alzheimer's true age range. In a 1968 milestone study, they reported that upon examining tissue from deceased demented elderly patients, 62 percent had the very same lesions that had scarred Auguste D.'s brain.

Alzheimer's disease, in short, was no rarity! It was a stunning revelation, one that fit with the increasing presence of dementia that medical practitioners were observing among the elderly. Late-onset Alzheimer's had remained disguised as long as the population's average life expectancy had kept it under wraps. At the start of the century, while some people survived to eighty or beyond, the average life expectancy at birth was only forty-nine. (Alois Alzheimer, for instance, died in 1915 at age fifty-one.) Even by 1950, people in this country on average only lived to sixty-eight. In other words, in the past most older people didn't get Alzheimer's; before it had a chance to take hold of them, they had died of something else.

By 1970, the average life expectancy had nudged up to age 70 in the United States, by all accounts explaining the sharply rising rate of dementia. Those physicians who heeded Tomlinson, Blessed, and Roth's discovery that Alzheimer's could occur in later life now knew they weren't dealing simply with "insanity" or "senility." These older terms no longer held water. But the habit of dismissing a lost mind as an inevitability of aging was so ingrained that some practitioners never adjusted their thinking. They continued to cast off dementia as a hazard of aging, when, although Alzheimer's type dementia was untreatable, they might have cushioned its blow through empathy, counseling, or drugs to reduce a patient's agitation. To this day, Alzheimer's as well as other causes of dementia in older people still slip by, both in the doctor's office and the general community. Observes geriatric psychologist Paul Raia, assistant director of the Massachusetts Alzheimer's Association, "Someone will tell their doctor that their seventy-five-year-old father is getting lost in his car, and the doctor will respond, 'Oh, he's just getting old!'" There's still plenty of confusion, says Raia, between the ordinary deficits of growing older and those derived from a disease.

Once Alzheimer's breadth of devastation was recognized in the early 1970s, the motivation to understand its underlying causes increased substantially. But observations by electron microscopists could go only so far in getting at the disease's underbelly. Fortunately a medical breakthrough in the late 1950s had other neuroscientists taking a different track. Parkinson's disease and its decline in movement and speech had been linked to a major loss of neurons in the substantia nigra, a small midbrain region that generates the neurochemical dopamine. Dopamine supports another region higher up in the brain, the corpus striatum, the neurons of which are critical to movement. Dopamine-deprived, the corpus striatum lost function and a person lost normal mobility.

This forced the question—could Alzheimer's and other neurological ills be tied to similar declines in certain brain chemicals, or neurotransmitters? There were plenty to investigate. (Presently, some fifty neurotransmitters have been identified.) A few research groups decided to inspect acetylcholine, a brain chemical produced by cholinergic neurons found primarily in the nucleus basalis, a small region in the base of the forebrain. It was imagined that acetylcholine might support memory regions in other parts of the brain, since a drug that blocked its action had been shown to impair memory. In 1976, three British teams sealed the data. They found that levels of acetylcholine plummeted 50 to 90 percent in Alzheimer patients, depending on the stage of the disease, while other measured neurotransmitters stayed within normal ranges.

It became more evident that neurons in brain areas hard hit by Alzheimer's use acetylcholine. Very believably its deficiency was an important contributor to the illness. Thus emerged "the cholinergic hypothesis"—notable especially in that it was the first time scientists had pulled a clue about the disease directly from its chemical abyss. Parkinson's researchers had come up with L-dopa, a drug that the brain converts to dopamine. The hope was that drugs might similarly compensate for acetylcholine in Alzheimer patients. "It was the beginning of rational treatments," recounts Peter Davies, a neurochemist who helped clinch the cholinergic evidence.

The year 1976 went far in creating a strong vine of effort against Alzheimer's, both at the scientific level and on the broader national front. Appearing that year in the *Archives of Neurology*, an eloquent editorial by Robert Katzman, chairman of Albert Einstein's neurology department, dramatically underscored the disease's alarming status as "the fourth or fifth most common cause of death in the United States," and how certain practices continued to cloak its rampancy, as when coroners overlooked the presence of Alzheimer's and instead attributed death to pneumonia, dehydration, or another secondary condition brought on by Alzheimer's. Katzman, his colleague Robert Terry, and a growing circle of others, would work hard to dispel the notion that Alzheimer's was an oddity. Soon a conference of scientists met at NIH to set an agenda for Alzheimer's research, and two years later another NIH gathering paved the way for the 1980 establishment of the Alzheimer's Disease and Related Disorders Association. ADRDA's name would shorten to the Alzheimer's Association.

Throughout the United States, families with dementia in their midst were banding together in search of more accurate diagnoses, treatment measures, caregiving facilities, and medical coverage, not to mention sympathetic advice about their ordeal. Their cries for help usually fell on deaf ears. So protracted was the disease—lasting from onset to death an average of eight years, and frequently much longer—that a husband or wife, daughter or son often ended up the providers of exhausting year-after-year care that trapped them in a "living hell," as George Glenner called it. When "Desperate in N.Y." wrote a short letter to columnist Abigail Van Buren in October 1980 seeking advice about her husband's probable Alzheimer's, "Dear Abby's" reply suggested the woman contact ADRDA. Reading this brief exchange in their daily newspaper, 30,000 other people deluged ADRDA's modest Manhattan office over the following weeks with mailed inquiries about the disease. "You are not alone," Van Buren had conveyed to the woman, and indeed she wasn't.

Americans learned a year later that Hollywood's finest—Rita Hayworth, Hollywood's sultry "Love Goddess"—had Alzheimer's. It both shocked a nation and explained Hayworth's erratic behavior in recent years. In a society that values youth and places a premium on sharp, bright minds (in what some today call a "hypercognitive" nation), her affliction deepened the reality of Alzheimer's awfulness. When Hayworth died in 1987 at age sixty-eight, President Ronald Reagan, her frequent costar, observed, "Her courage and candor and that of her family were a great public service in bringing worldwide attention to the disease which we all hope will soon be cured." Hayworth's daughter, Princess Yasmin Aga Khan, has maintained a vigilance ever since by holding annual fundraisers to benefit the Alzheimer's Association.

By the time George Glenner turned his focus to Alzheimer's amyloid in the late 1970s, the disease had leapt from near obscurity to being regarded as an epidemic. An astonishing 100,000 Americans were thought to lose their lives to it annually. Two million were reportedly afflicted with it. Born under the sign of Virgo, it's as though Glenner's tendency to never be so much as a whisker late for an appointment guided him into Alzheimer research precisely when the disease's stunning prevalence and the graying of society were inciting a mood of retaliation. With or without Glenner, science was readying to do battle with a villain.

Had the disorder still been considered a rarity would Glenner and, in time, so many others of us in the scientific community have pursued Alzheimer's so single-mindedly? "When something starts to move, visionary scientists like to be where it's hot—where the excitement is—where the problems are tough and nobody has succeeded," remarks William Comer, a prominent drug researcher who enters this story later on.

There's little doubt that Glenner wanted to be where it was hot, where a research commitment might transport millions of people out of a bleak fate. When he and his wife, Joy, moved to the West Coast in 1980, Glenner saw his new post in the Department of Pathology at the University of California, San Diego, as an invigorating fresh start. Upon its opening in the early '60s, UCSD had been the first American university to establish a department of neuroscience as well as an attached doctoral program. This precocious attention to a field whose day was just dawning meant that Glenner was working in as resourceful an environment as a brain researcher could ask for.

Once he arrived on the La Jolla campus, Glenner was considered by some to be the man who knew most about amyloids and their underlying proteins. Yet when word got around that he was trying to isolate amyloid from the brain for the purpose of throwing light onto the cause of Alzheimer's, there were many doubters. "I greatly admired Glenner; he was a maverick and knew his stuff. But people weren't even sure he was on the right track—or that there was a protein to be found that caused this mess in the brain. I myself believed he only had a fifty-fifty chance," notes Vito Quaranta, a cell biologist at Scripps Research Institute in La Jolla and a friend of Glenner's. Even though amyloid had been identified as a major component of Alzheimer's brain-strewn debris way back in the 1920s, it might have zero bearing on the disease's origins. If something about the plaques did play a central role, it was known that dozens of other proteins lay embedded in these clusters, so amyloid might not be the worst of the lot. Or maybe a protein that had nothing to do with the plaques was the instigator—one connected to the tangles, some guessed.

Paying not much heed, Glenner went his own way. Tall and lean, with wavy silvering hair, bushy eyebrows, and a long nose, he was called the "silver fox" by one of his friends, and indeed he chased the details of his work with a fox's eye. All told, he was busily embarked on a dual mission, for he was attempting both to solve Alzheimer's and, as will be seen, to care for those stricken by it. Keeping to a strict self-imposed agenda, as a researcher he tended to be on the aloof side. He didn't collaborate all that frequently with other labs, and, of the older school, rather than turn the lab work over to a posse of postdoctoral students, he preferred getting his own hands dirty at the bench, his self-reliance extending to his occasional blowing of glass to taper tubing. His hearing impairment is thought to have contributed to his isolation and, in all likelihood, made his eye that much more alert.

As consumed as he could be, his friends and close colleagues recall a warm, good-natured man of gracious, formal manners and a quiet reserve, whose cashmere cardigans and tweed jackets with suede elbow patches—the result of frequent excursions to Nordstrom's—bespoke "gentleman." (His wife, Joy, recounts that when she first acquainted Glenner in 1976, "I'd never met such a handsome man who dressed so terribly.") Out from under this formal exterior, a playfulness could emerge. Glenner's infectious grin and a certain twinkle in his eye would signal that he was up to something. On other occasions, this consummate scientist—who considered lunch an interruption—could be gruff and territorial toward competing researchers. He had reason to be, says Joy. Years before, he'd felt badly betrayed by a close colleague who had reviewed his work for a journal, then published the same material ahead of him. Yet, deep down, according to Joy, Glenner was really "a pussycat."

"The proof of the pudding," as Glenner saw it, just required perseverance and patience. Even if he was wrong and amyloid wasn't central to Alzheimer's, that in itself would be an important finding. In its purest form, scientific method is intended to disprove existing hypotheses in order to get closer to the truth. But if amyloid *was* a central player, "there's going to be hellzapoppin!" he told Joy, with whom he shared every detail from the lab. For starters, knowing the chemical makeup of Alzheimer's amyloid, Glenner believed, might make it possible to easily diagnose Alzheimer's with a blood test; if levels of the protein were elevated, it might signal the disease. So far the only way to unequivocally tell if someone living had the disorder was to drill out a small circle of skull and remove a sample of gray matter for an analysis of plaque and tangle amounts.

Today, this remains the sole route to a definite diagnosis in a living patient. It's always struck me that someone burdened by dementia needs this added trauma about as much as a hole in the head, which is what the procedure essentially amounts to. Most physicians being of the same mind, a brain biopsy isn't usually employed. Moreover, it can miss lesion-replete tissue, thus delivering a false negative diagnosis. Physicians instead arrive at "probable" Alzheimer's through neurological and cognitive assessments and by ruling out other conditions that can lead to dementia. Alzheimer's and vascular dementia (the latter often due to stroke) together account for close to 90 percent of all dementia cases; Alzheimer's, alone, for roughly 70 percent. Still, a fairly long list of other dementia-causing situations has to be checked out—Pick's, Parkinson's, and other neurodegenerative disorders, autoimmune ailments, infectious diseases such as meningitis and AIDS, as well as brain tumors and head injuries, metabolic disorders, toxins, drugs and alcohol, and depression, to name but a few. In the early '80s, if "probable" Alzheimer's was arrived at, there remained a 10 to 30 percent chance of misdiagnosis.

A definitive diagnosis *could* be sought after death if a family so desired. Autopsied tissue could divulge the true status of plaques and tangles from tissue retrieved from two or more brain regions commonly hard hit by Alzheimer's. For instance, if the hippocampal and cortical tissue of a seventy-five-year-old demented individual contained fifteen or more plaques per square millimeter, it usually was clear evidence of Alzheimer's, as was a hippocampus overrun by tangles. But what good was a postmortem diagnosis to a patient?

As Glenner was so fully aware, even though Alzheimer's was untreatable, if you could diagnose it in the living, you at least might provide therapies that could reduce agitation and make arrangements for supportive counseling. Moreover, a blood-derived diagnosis could guard against confusing Alzheimer's with a treatable dementia, plus do away with the chore of excluding other conditions. And when a treatment arrived someday, physicians would have a means of knowing exactly who to treat.

But a diagnostic test was merely at the short end of what might result from apprehending the amyloid protein's sequence of amino acids. At the far end, Glenner envisioned a useful drug for Alzheimer's. He even dared occasionally mention the word "cure." Decoding the protein's structure might be like finding the key that unlocks a huge outer door that leads to many smaller inner doors and hidden chambers. It conceivably could bring to light a mutation in the protein's corresponding gene, and, for all anyone knew, other disease-related defective genes and proteins. Glenner was inclined to believe that environmental stressors somehow fanned the disease process. Yet if just one specific in-body root abnormality could be exposed, it might be drug-targeted, thereby slowing, halting, or, for anyone knew, reversing neuronal disarray.

Glenner and Wong's isolation of brain amyloid in June 1983, marked by the vivid apple-green seen through a polarizing microscope, had relied on what Glenner viewed as a scientist's greatest asset—imagination and "the ability to not be bound by fixed concepts. Being able to work around corners before you turn them." Doing just that, he had gone after the amyloid deposited on the walls of the brain's blood vessels, which had been shown to be more accessible than gray-matter amyloid. Accordingly, over a period of many months, Glenner and Wong had perfected methods for stripping off the meninges, the brain's outer protective layers that are interlaced with blood vessels, grinding them up, and weeding out the vessels' bits and pieces of connective tissue so they'd be left with only amyloid. The means they adopted weren't necessarily new, but in several instances they applied them in a special manner in order to squeeze out results.

As Cai'ne Wong remembers, "When George was enthusiastic, I wasn't, and when I was, George wasn't, which was something we did to keep ourselves in check." Wong, who'd recently gotten a degree in sociology at UCSD, had picked up an interest in proteins while earning money washing glassware in a biochemist's laboratory. When the job with Glenner had presented itself, he'd easily opted to put sociology on a back burner.

Handily, the brain tissue the two scientists worked over came from a brain bank that took up one corner of Glenner's three-room laboratory. Started by Glenner in the spring of 1981, it was one of the first such depositories in the world dedicated to Alzheimer research and comprised, quite simply, two enormous horizontal Revco freezers. If he was to find the molecular "answer" to Alzheimer's, as he constantly referred to it, Glenner felt it was imperative to have immediate, constant access to fresh tissue. Each Revco held up to 200 brains, and by 1983 the bank was more than half full. "It was one of the best advantages we had," relates Wong. "If you have no field, you can't be much of a farmer." A healthy brain usually weighed about three pounds—this even being true of sound older brains, which seldom lost more than 10 percent of their mass. In comparison, a brain wizened by Alzheimer's might be reduced to two pounds or less. Some of those sliced and examined by the scientists were so bereft of cellular volume that their sulci, the grooves along their surface, had deepened into gaping crevices.

Families contacted Glenner from near and far for a postmortem diagnosis of a relative's brain disorder. Glenner made arrangements for a prompt autopsy, after which tissue slices were prepared for diagnostic and research purposes before the organ was wrapped in thick plastic and stored in a freezer. For next of kin, an autopsy-based diagnosis could provide helpful closure. They might finally learn whether the fatal undoing really had been Alzheimer's—or possibly Pick's disease, limbic encephalitis, or some other dementing condition. Contributing a brain also provided a positive sense of striking back at the disease, not only through Glenner's research, but through the work of the many scientists nationally and internationally who requested samples from the brain bank.

But on occasion a diagnosis of Alzheimer's in a deceased middle-aged person could come to haunt family members. It could raise the fear that they, too, were genetically doomed. There'd been a long-running debate in the medical world over whether Alzheimer's could be inherited, and increasing reports of families with several early-onset cases in two or more consecutive generations were strengthening the opinion that Alzheimer's rare, early form probably was passed down by an autosomal-dominant mutant gene. Some researchers, however, still staunchly maintained that environmental hazards triggered the disease, not genetics. This disagreement left offspring and siblings of patients in confused limbo.

The next stage of Glenner and Wong's research had a tricky aspect to it. To deduce the amyloid protein's amino-acid composition, they would have to dissolve the brick-hard amyloid just enough to pull apart its chemical letters, but not too much or they'd degrade the substance and obscure its sequence. As the work began, another La Jolla summer showed every sign of following its typical course: "May gray, June gloom, July fry, August die." Several times a week Glenner left the lab's air-conditioning for the hot outdoors, walked down campus paths flagged by eucalyptus, caught the campus shuttle to the San Diego community of Hillcrest, then hoofed it to Third Avenue. There sat the small cottage, with pink geraniums in its window boxes, that he and Joy recently had turned into one of the country's first designated day-care centers for Alzheimer patients.

A singular event the previous summer had sparked the Glenners' nearly overnight creation of the Alzheimer's Family Center. After eating dinner at their La Jolla home, they'd gone to Glenner's laboratory, as they frequently did in the evening. Glenner always felt as though he had to keep pressing forward and liked Joy to come along to keep him company. Joy—Glenner's second wife—shared her husband's deep commitment to solving Alzheimer's, particularly as she'd watched her own mother suffer through years of stroke-induced dementia. In the lab she lent a hand with paperwork or washed glassware. At about one in the morning, just as they were about to call it a night, Glenner received a phone call from an elderly, distraught man: "I'm not going to give you my name, because I don't want you to stop what I'm going to do, but when the police get here they'll find a note that says to make sure that my wife's brain comes to you for your brain bank so we can stop this horrible disease. I'm going to kill us both."

While Glenner tried calming the man, Joy had the call traced. The caller was in Julian, an old gold-mining town an hour's drive east of La Jolla. Julian's sheriff was contacted, and when he reached the man's residence, he found the man with a gun in his hand, still on the phone with Glenner, and in the next room his wife, profoundly enfeebled by dementia. The following morning the Glenners drove to Julian to meet with the couple. The husband, in poor health himself, hadn't wanted to put his wife in a nursing home or burden his son with his problems. The son, once contacted, awoke to how dire the situation had become and immediately brought his parents to live with him and his wife.

"It was a very traumatic episode for us," relates Joy, "and driving home that evening, Glen turned to me and said, 'Hey kid'—he always called me kid–'what are you doing tomorrow?' I asked why. 'Because,' he said, 'it's time for us to start taking care of the living.' The very next day he started looking for a place to rent. We found the Hillcrest cottage within two weeks, and within six weeks we had obtained start-up funding, a license, a board of directors, and nonprofit status, and with the help of volunteers had completely renovated the cottage, which had been falling apart."

Their new facility's opening gala on October 2, 1982, hadn't turned out quite the way the Glenners planned. After setting the date, they'd received an invitation from the White House for a ceremony scheduled the very same day, the occasion being President Reagan's signing of a proclamation citing Thanksgiving week as National Alzheimer's Disease Awareness Week. Glenner—a former top pathologist at the NIH and, by then, a scientific adviser to the Alzheimer's Disease and Related Disorders Association as well as a cofounder of San Diego's local Alzheimer chapter—was an obvious invitee. Moreover, recounts Joy, it was the Glenners who had proposed to California congressman William Lowery that Thanksgiving week be set aside for Alzheimer's, Lowery having created the legislation for Congress's approval. "It was something we couldn't not do," remembers Joy, so they packed their finest and flew east for the Oval Office gathering, leaving others in charge of their center's launch party.

Years later, George Glenner would tell a San Diego journalist about his exchange that October day with President Reagan. "The President looked at me and asked, 'What is Alzheimer's disease?' I explained that it was tangles and plaques that form in gray matter that keep neuronal cells from being nourished. Well, he looked at me and smiled. 'All I know is that my mother died in a nursing home and she didn't recognize me at the end.'" Glenner had his photograph taken with Reagan, who gave his sporting, assured smile.

For Glenner, visits to Hillcrest were a welcome change from the laboratory's often frustratingly slow pace. Ever since the five-room day-care facility's door had swung open to six patients, the progress had been immediate and gratifying. By July 1983 there were twenty patients, and numerous referral calls were received daily. Staffed by Joy—the center's CEO and program coordinator—along with two nurses and a handful of volunteers, the Alzheimer's Family Center kept at bay the "living hell," as Glenner saw it, that family members had to endure every bit as much as a patient. Glenner, the day-care's medical adviser, found that witnessing the disease's ruthless course in his Hillcrest patients strengthened his determination in the laboratory. At the same time, his compassion for patients and their families was all the greater because of the disturbing brain clutter of plaques and tangles he observed under the microscope.

"Glenner was on fire about helping his patients," recalls Dennis Selkoe, who got to know Glenner in the mid-'80s when they both served on ADRDA's scientific advisory board. "He talked about his clinical work as passionately as he did his lab work. He wanted to do the perfect thing in medicine—go from the bedside right to the cure in the lab."

A textbook definition of Alzheimer's mental symptoms includes losses in four principle abilities—memory, orientation, judgment, and reasoning—and as the Hillcrest staff came to appreciate, these deficits encompass an endless sphere of broken connections between the patient and his surroundings. The failing mind takes with it the ability to find words and converse, to write a note, to read a book, to identify by touch a blade of grass or a knife, or recognize a drink by its taste, or realize that a passing object overhead is a bird or a cloud. Without a recalled context, vision and the other four senses become empty vesicles. Lost connections invite delusion. The dark interior of a closet can look like a frightening gaping hole. A train's whistle can sound like a bloodcurdling scream. A glance in the mirror, and a patient may not even recognize herself and contrive someone from out of her worst fears—her husband's mistress. Even the devil.

All the blackest emotions can rise up, including depression, anxiety, paranoia, panic attacks, and combativeness. Agitation and confusion can extend to hallucinations, delirium, sleeplessness, and a habit that caregivers called "sundowner syndrome," the tendency for patients to grow especially confused and disoriented late in the day and prone to wandering, particularly in unfamiliar settings. This behavior, according to Joy Glenner, most likely occurs because by day's end a patient's hard-pressed ability to cope is all the more diminished by fatigue.

Given their plight, one might imagine that many patients would choose suicide. Yet in comparison to those trapped by certain other incurable neurological disorders—Huntington's disease, for example—few Alzheimer patients follow this course, and it's not solely because the disease (at least not in its earlier stages) makes them forget their illness. In Huntington's, while a person's motor control declines early on, their mind usually doesn't, so they retain insight into what their worsening condition means for their future, observes Marilyn Albert, Mass General's director of Gerontology Research and an HMS professor of psychiatry and neurology. In contrast, Alzheimer's disease can reduce the capacity for insight at a relatively early stage. "Alzheimer patients have trouble projecting into the future and seeing the disease's long-term consequences. You have to have insight into what's happening to you in order to get depressed and want to commit suicide," says Albert. Those afflicted may even deny they have memory problems, and here again, says Albert, it's not necessarily that they're in denial. Instead, they may not be able to make the connections that allow them to perceive the changes they're undergoing.

As a common saying among health professionals goes, it's not the people who say, "Oh no, I'm forgetting things, I must have Alzheimer's!" who get the disease, but those who forget things and don't say anything.

Glenner always was struck by patients who, in the early stages, spoke so knowledgeably and fluently, and could seem so mentally fit. They concealed their inability to process and store new memories by drawing from their deep well of long-held memories. There was Martha, for instance, who dressed immaculately—on her own, her daughter said—and talked in detail about the many years she taught history. But whenever she spoke of her recently departed husband as if he were still alive, Glenner felt he was looking straight into the jowls of the disease. In later stages, even after a person's cognitive abilities plummeted, their social graces could shine through in surprising ways, as if one held on to these symbols of human dignity beyond all else. "Manners are one of the first things we learn," notes Joy Glenner. "They're long-term memory and one of the last things a failing mind loses."

One day at Hillcrest, Mary, one of the nurses, crouched down to plug in the television, lost her balance, and fell over. Three male patients, all with advanced Alzheimer's, jumped up from their chairs and rushed over as best they could to help her up. Mary and other staffers couldn't help but be amazed over this gentlemanly stampede. It seemed so incongruous, given that the three patients no longer could use a telephone, drive to the market, or make or even eat a sandwich unassisted.

In the early 1980s, hardly any manuals existed to tell the Glenners how best to care for a person with dementia. However, with input from gerontologists and caregivers, they put into place a daily schedule that kept the patients as active and oriented as possible: aerobic exercises and fresh-air walks, arts and crafts, music therapy, popcorn breaks, massages and quiet time, story-telling, the reading aloud of newspapers ("reality orientation"), and endless games—ring toss, volleyball, miniature golf, bingo, horseshoes. Music hour often was in progress when Glenner arrived in the afternoon. Straight-backed, eighty-four-year-old Lydia would be seated at the upright Baldwin, which had been donated by the Lawrence Welk Foundation, hammering out tunes from the Fabulous Forties. The music brought many of the patients back to a fondly remembered time. A tribute to her still solid long-term memory, Lydia's nimble fingers never missed a note, although, on the other hand, she could forget she had played a song and might play it over several times, until, like correcting a skip on a record, someone would tap her on the shoulder and she'd start in on a new tune. Glenner, whom the patients called "Dr. G.," danced with one patient after another, covering all the ladies. Engaging socially with the patients, he felt, was as important as responding to their medical needs. As he danced them round, he bent his six-foot-two stature down to make eye contact. This was the best way, he believed, of getting a patient's attention and connecting.

Gradually the staff was learning the do's and don'ts. Do reinforce a patient's strengths. Tell Alice how much you like her sculpture, even if you're not sure what she's making. Don't foist reality upon a patient. If Henry thought he was in his boyhood town of Rosemont, so be it. If Sam thought someone had stolen his wallet, commiserate with him. Do offer a range of choices—not in the form of a question, which lent confusion and could throw a patient into a catastrophic reaction, but by redirecting them. If they didn't like going for a walk, then suggest a craft or a game. One afternoon at Hillcrest an unaware volunteer asked Walter, a seventy-seven-year-old who was six years into the disease, who the lucky person was who would receive the greeting card he was coloring. A stricken look contorted his face. He mumbled loudly, mouthing a long string of vowels void of consonants. Gauging by the apprehensive look in his eyes, he might have been trying to say, "I understand what you're asking, but I'm unable to reply, and I feel awfully about that." The volunteer quickly learned to appreciate not only what a patient couldn't respond to, but also the huge effort it required for them to respond in the first place.

Walter's walk was a hesitant shuffle. The disease had traveled far enough to have begun its attack on his brain's centers of motor control. In time, his cognitive and motor deficiencies would blend into a general incapacity. He might forget how to, or simply be physically unable to, walk or hold up his head. Already he could no longer smile. Alois Alzheimer's patient Auguste D. had ended her days curled like a fetus, the position the body instinctively remembers best. There was a good chance that Walter, like a significant portion of Alzheimer patients, would die from the secondary impact of pneumonia. When the ability to swallow is forgotten or lost, food and liquid can become aspirated in the lung, lodging there and triggering infection.

The Hillcrest staff increasingly realized that the best they could do for their patients was to try to remove whatever stress, whatever decisions, whatever unknowns might weigh on a patient's fragile mind. The end goal of their care became supplying security and, above all, love. "We discovered there really is a treatment for Alzheimer's disease," says Joy Glenner. "It's a glue called love. You can have a very disoriented or combative person, and love can help calm them." Notes psychologist Paul Raia, "Normally our everyday reality relies heavily on our memory and learning channels into the brain, but when Alzheimer's disease shuts down those channels, a person's reality becomes rooted in what they see and they feel. Their emotional experience becomes their reality. Therefore, maintaining positive emotions is the caregiver's first challenge."

A daughter of one Hillcrest patient once called Joy to ask why her mother returned home from day-care each day with stains all over the back of her blouses. Joy and the rest of the staff thought and thought, until they suddenly realized: the stains were from hands, all the hands of other patients and also staff members who touched, hugged, and gave comfort to one another throughout the course of the day.

La Jolla's July temperatures soared as Glenner and Wong prepared a strategy for solubilizing the amyloid they had in hand. How anticlimactic if they'd gotten this far only to be unable to read the protein's amino-acid script. Glenner realized that the beta-pleated sheet structure of amyloid proteins, which he'd discovered they all had in common, made them nearly impervious to insult. You could put amyloid in water and mix in the most efficient solvents and detergents—even the notoriously harsh SDS (sodium dodecyl sulfate)—and amyloid meanly sat there, resistant to dissolving. To enhance amyloid's solubility, Glenner's plan was to disrupt water's structure as well as certain chemical aspects of amyloid. For this feat, which is commonly resorted to in biochemistry, the scientists immersed the amyloid in water containing a chaotrope, a specialized salt.

Slogging forward they had made it to a higher station. But here they hit a surprise fork in the road. When they ran their amyloid solution over a separating column, they discovered they didn't have just one protein suspended in solution, but two. Both were very short sequences. Genes don't usually make such short proteins, and they suspected right away that each of their molecules was a peptide—a piece of a protein. Which of the two constituted the amyloid gunk that clung to the walls of cerebral vessels wasn't obvious, so they decided to go with the second one they'd found, and thus named it "beta"—the second letter in the Greek alphabet. They chose it simply because they had isolated more of it, according to Wong, and meanwhile put the peptide they named "alpha" on hold.

They remained far from confident, however, that their *beta protein* was Alzheimer amyloid's main component—the primary stuff of both the cerebrovascular and gray-matter deposits. If it withstood further analysis, "beta" would be an apt name for it, given the beta-sheet composition of other bodily amyloids.

The day they set out to read their peptide's sequence, neither Glenner nor Wong dared be optimistic, for they might find that its units were insurmountably unreadable. "The chain of amino acids we were tackling was like a string of beads, with each bead being a different color," describes Wong. "You had to advance bead by bead." Only when the first bead was pulled off and its color, or chemical, identified could the color of the next bead be identified, and so on. The job was left to a Beckman amino-acid analyzer. The first bead identified—aspartic acid—led to the identity of the second bead—alanine, and bead by bead the process continued, until by the end of the day, by what seemed a miracle, the Beckman had cracked the peptide's sequence! The revealed stretch amounted to only twenty-four amino acids. "George got home really late that night," remembers Joy Glenner, "and the first thing he did was to come upstairs and poke his nose around the corner of the bedroom door. 'Break out the champagne,' he said. 'We've got it! We've got it!'"

Glenner and Wong's report of their breakthrough, entitled "Alzheimer's Disease: Initial Report of the Purification and Characterization of a Novel Cerebrovascular Amyloid Protein," wouldn't be published until the following spring, when, with little fanfare, it appeared in the May 16, 1984, issue of the journal *Biochemical and Biophysical Research Communications (BBRC)*. Glenner, according to Wong, never submitted the paper to the prestigious journal *Science*, because years before its editors had rejected a paper he had deemed especially publishable. He also found *Science*'s peer reviewers—a paper first having to gain the approval of other researchers—too narrow a bunch. He felt he'd be throwing food to his shark competitors.

Others who had attempted to isolate Alzheimer's amyloid from gray-matter plaques had run aground, largely because plaque amyloid is less soluble than the amyloid that lines the brain's blood vessels. Glenner's choice "to go where the light was brightest," in Wong's words, had paid off.

A big unknown remained. Had the scientists sequenced the *right* protein—the primary substance of both Alzheimer's blood-vessel and gray-matter deposits? Glenner turned to his friend Vito Quaranta at Scripps, who, in collaboration, raised antibodies to the beta peptide, which—hallelujah—recognized the accumulated debris in both sites, but not the tangles. Glenner and Wong could feel confident that their isolated beta fragment represented the plaques' core material. Around the same time in 1985, a team headed by Colin Masters in Australia and Konrad Beyreuther in Germany went one important step further. Isolating amyloid directly from plaque cores in brains affected by Alzheimer's and Down syndrome—another disorder that heaped amyloid upon the brain—they retrieved a peptide similar to Glenner's. This provided all the more confirmation that Glenner had retrieved the molecule he was after.

Glenner and Wong's May 1984 paper in *BBRC* got scant mention by the press. Yet its huge ramifications weren't lost on the scientific community. Nearly eighty years after Alois Alzheimer's sighting of peculiar matter in Auguste D.'s cerebral cortex, scientists had its chemical identity in hand.

Right up until Julia was thirty-nine and pregnant with her tenth child, she was nonstop energy and cheer. Always immersed in household chores and shuttling her children around town, she was hardly ever off her feet. Throughout the day she sang like a lark, so much so that her children grew up thinking that all mothers sang incessantly while pinning clothes to the clothesline or ironing or baking. Her father having been a baker, his skills had rubbed off on her, and the Noonans' busy house always smelled of cooking loaves, biscuits, or notoriously rich brownies known to every child in the neighborhood, not to mention Julia's Sunday night treat—fried dough. Children were her passion. She loved having them around, whether her own flock, nieces and nephews, her children's classmates, or neighborhood strays, and every night there'd be one or two extra heads at the dinner table. If, of late, she occasionally misplaced a glove or the car keys, or forgot a dentist appointment, such lapses were seen as perfectly normal for a mother who had her hands full.

— *three* —

Candidate Chromosome

There's something happening here,
And what it is ain't exactly clear.
—Stephen Stills, "For What It's Worth"

In the early 1980s, the chromosomes of people, zebras, potatoes, ants, pine trees, and all other living things lay as physically unexplored as the distant planets. Scientists had photographed these spiraling DNA strands, but few on-site missions had charted their terrain in depth. With respect to the human genome, twenty-three pairs of chromosomes lay packed inside the nucleus of most every cell—each chromosome an impossibly long corridor of genes (which encode proteins) as well as intervening stretches of "junk" DNA (which don't). Over 95 percent of our DNA, in fact, is composed of noncoding stretches that occur both inside and outside genes. Whatever purpose this wasteland might serve remains largely undetermined. Some areas may be used in gene regulation, others to provide structure for the chromosome.

Naturally, the biggest problem about exploring chromosomes and isolating their genes was DNA's minuscule size. A DNA strand from whatever living species is a mere 60 trillionths of an inch thick. You can't exactly sit down and read off its four bases—adenine (A), thymine (T), guanine (G), and cytosine (C)—the way you can the lettered contents of *War and Peace*. (If written out, the human genome's 3 billion letters would require nearly 1,000 volumes, each as long as Tolstoy's classic.) But if researchers were to become privy to genetic explanations for how the eagle or the ladybug develops wings, or why some people can curl their tongues and others can't, or how it is that carp have fifty-two pairs

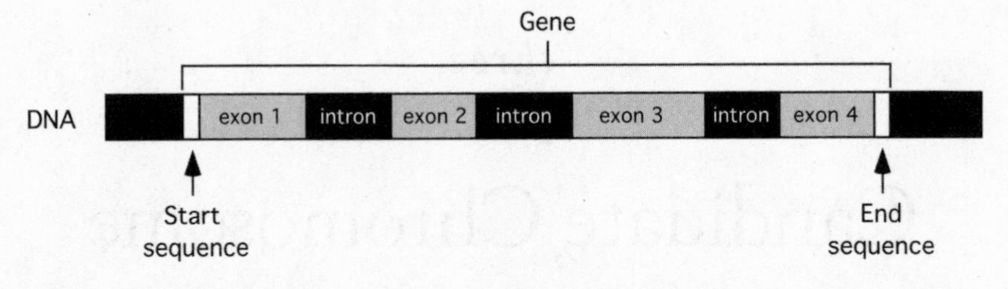

FIGURE 3.1 A gene and its intervening stretches of coding (exon) and non-coding (intron) DNA along a chromosome. Illustration: Robert D. Moir

of chromosomes compared to a human's twenty-three or a tomato's twelve, or exactly what separates a firefly from a carrot, or why some people get Alzheimer's disease at forty while others are able to master the computer at eighty—if all sorts of questions were to be answered about how life at the molecular level works, or doesn't work—the DNA book as it pertains to each species had to be read cover to cover.

A carrot and a firefly, spun from the same DNA fabric, share more at the molecular level than meets the eye. Consider how scientists have been able to implant a fruit fly's "glow" gene into a carrot, producing a glowing carrot. So interchangeable among organisms are life's basic chemicals, it's a wonder—and, if we meddle too much, potentially a danger. Nature has fine-tuned an organism's DNA specifications over so many eons, we're unlikely to ever fully appreciate the natural wisdom behind the design. If here and there, there and here, we insert and delete too many genes, we could end up like the sailor who, just when he's lulled into thinking he knows the sea, gets into trouble. The sum of the parts is so staggeringly, unknowably huge.

In the midst of the Huntington's project, Jim Gusella had tossed an interesting idea my way. Gordon Stewart—a new research fellow in Gusella's lab—myself, and Paul Watkins, who was working at the biotech company Integrated Genetics, were continuing to turn up DNA variations on chromosome 21 in our effort to find a marker that linked to Huntington's. Using those markers, why didn't I start plotting a map of chromosome 21? So far, no DNA variations from 21 had hooked Huntington's gene. As we'd discover, they wouldn't, of course, since Huntington's defect instead lay on chromosome 4. But we were accumulating enough chromosome 21 variations, Gusella suggested in his insightful way, that mapping them might shed light on Down syndrome.

Down's and its erroneous *three* copies of some or all of chromosome 21 instead of the normal two copies produces a wide constellation of features—from facial changes, to heart and lung problems, to mental limitations. Figuring out where the DNA variations we had isolated sat relative to each other along chromosome 21 might help identify which triplicated parts of the chromosome correspond to which Down's features. We might find, for instance, that one triplicated portion of the chromosome—and therefore a triplicated gene therein—leads to heart defects, or another to skeletal abnormalities.

Unbeknownst to us, in California, as George Glenner was making ready to isolate amyloid from Alzheimer-racked brains, he too was intrigued by Down syndrome and its tie to chromosome 21, but for a different reason. Apart from Alzheimer's, Down syndrome was the only other condition known to produce an appreciable excess of amyloid in both the brain's gray matter and its blood vessels. Glenner had to wonder, as did others: Because both these diseases resulted in so much brain amyloid, could this mean that Alzheimer's, like Down's, arose from a chromosome 21 abnormality? He and his first wife had three children, one of whom had been born with Down syndrome. He seldom spoke of her in the lab, yet she was forever in the back of his mind. What if the amyloid proteins in Down's and Alzheimer's were chemically identical? Would that mean that on top of her present disability, she'd later fall victim to Alzheimer's?

At Mass General, I was delighted with the notion of trailblazing a human chromosome. John Wesley Powell and Henry Hudson explored and navigated wider waters, but none so descriptive or prophetic of their own species. In existence were partial linkage maps for chromosomes of various bacteria, but hardly any detailed maps existed for human chromosomes. With an eye toward our Huntington's gene search, countless other groups also had begun looking for variations in DNA that might link to disease genes, and you could sense the human genome's enormous dark continent very gradually opening to the light of day. Maps of markers along chromosomes would soon be in high demand.

Chromosome 21 is the shortest human chromosome, so the idea of mapping it was all the more feasible. A human's twenty-three pairs of chromosomes originally were numbered according to size—chromosome 1 being the longest. But along the way someone mismeasured, because 21 is actually shorter than 22, although the original numbering was left to stand. (The sizes of the twenty-third pair—the sex chromosomes, in which an XX pairing, as symbolized, begets a female, while an XY pairing begets a male—are an exception. The X chromosome is about as large as chromosome 6; the Y, nearly as small as 21.) It's mainly due to chromosome 21's small size that children with Down syndrome survive at birth. Larger chromosomes, if erroneously duplicated, often result in miscarriage because the fetus isn't able to endure the excess activity of so many extra genes and encoded proteins.

Time in the laboratory was consumed by the Huntington's project, so I took to building the chromosome 21 map on the daily back-and-forth bus ride between Providence and Boston. I was using chromosome 21 variations primarily found in the Venezuelans. Essentially it was like doing an endless crossword puzzle. Through mathematical calculations and visual assessment of the Venezuelans' family tree, I was following the inheritance of variation markers and filling in the blanks. Sitting in the back of the bus, family trees and autoradiographs piled on my lap, I manically crunched numbers on my pocket calculator, then scribbled them down, raising the curiosity of fellow commuters. They'd become even more curious when I told them I was mapping a human chromosome. "You're doing *what?*" I'd try to show them. "You watch to see whether tiny DNA variations found on an individual's pair of chromosomes remain together or stray apart in descending generations of one family. How often they get inherited together gets translated into genetic distance measured in centimorgans. One centimorgan roughly equals 1 million bases of DNA, so I can slowly piece together how close or far apart various sites are on the chromosome."

If the fellow commuter's eyes hadn't glazed over, I threw in more detail. "Say you've got two parents with twelve kids," as can be the case in a Venezuelan family. "Since each parent has two chromosomes, for each child there are four potential segregations of DNA that can occur from parents to child; so for twelve children, potentially you have forty-eight segregations of parental DNA. Say you follow one parent's two chromosomes, and you see that out of twenty-four segregation events, two of the variation markers were separated by recombination events three times during meiotic exchanges, so that means you've observed three out of twenty-four—roughly 12 percent—recombination events between these two markers, and that translates into twelve centimorgans of genetic distance between these two markers, or roughly 12 million bases of DNA."

After listening to this explanation, most people politely vanished back behind their newspaper. But a few commuters who shared the same bus schedule really got into the map's progress. "Hey, Rudy, how's the map going?" I'd give them an update. "The southern tip is just about done." Speeding me along, I admit, was the knowledge that a team at Johns Hopkins was working on a similar map. Ultimately they didn't get much further than plotting the position of one gene on chromosome 21 along with a few of the markers that Paul Watkins and I had dredged up for the Huntington's search.

Due to the snaring of the Huntington gene's approximate location on chromosome 4 in the summer of '83, our work on the chromosome 21 map slowed to a crawl. The temptation was to shout the Huntington's coup from the highest mountain, but the lab had to keep mum until there was stronger evidence of the gene's locale. Only then could the breakthrough be reported on. Keeping my lips sealed wasn't one of my strong points back then, especially as I was born into a talkative, very interactive Italian family. But the barnyard world of molecular science increasingly has taught me a fair amount about discretion, for if you tell the gander, the gander might tell the horse who might tell the goat who might tell the pig who might tell the rat, and if that happens you're really in trouble, because all the new data you worked so hard to collect aren't yours anymore. A lab wants to disseminate information as quickly as possible to push science along, but it first has to make sure the information is accurate and that it will garner due credit for the sake of procuring the next grant, or a lab can find itself out of business.

After the Huntington's triumph went public that fall, our humble advances into chromosome 21's linear maze continued. Those of us stitching the map together, a group that included several of Gusella's technicians, had reason to be proud. Our efforts were among the first to span the entire length of a human chromosome with markers—eventually thirteen. I distributed the map at conferences, and until more informative charts came along, it served as a Rand McNally for geneticists who wanted to travel onto chromosome 21, in most cases to rule in or out genetic linkage of a specific disease to chromosome 21. Upon request, collaborators received the map and the all-important tickets—DNA probes that bond to specific stretches on the chromosome that contain variations. Then their tests could take them "walking" and exploring for genes up and down the chromosome—"walking" being genetics vernacular for systematically extracting overlapping sections of DNA and progressing bit by bit along the chromosome in a northerly or southerly direction, often for the purpose of closing in on a gene. ("Sleepwalking," notes my colleague John Hardy at the Mayo Clinic in Jacksonville, is when you're walking along and bump into a gene other than the one you're looking for.)

Mass General investigators would use the map's probes to look for mutations in chromosome 21 genes associated with the movement disorder torsion dystonia, Lou Gehrig's disease, and other sicknesses. And eventually our chartings even achieved the original goal of correlating certain regions on 21 with specific symptoms of Down syndrome.

The chromosome 21 map wouldn't reach print until 1985—and then only in an obscure book on chromosome 21's molecular structure. Its formal publication in a journal didn't happen until 1988. The delay mostly was because Watkins, Stewart, and I got sidetracked by strange sightings on chromosome 21. As was already established, 21 wasn't your classic chromosome in which a shorter arm lies above a longer arm. It had a long arm, but its short arm resembled a squashed hat. A few other chromosomes have similarly dwarfed short arms. This breed is referred to as "acrocentric," because its center—where the short and long arms meet—is high *(acro)* up on the chromosome. We became increasingly aware that chromosome 21's squashed hat contained all sorts of DNA weirdness—multiple repeats and odd inversions of bases. Whatever genes lay in this upper region, we couldn't help but think, were rife with all sorts of disease-related mutations deserving of investigation. In hindsight, this proved not to be true. Mutations on the chromosome's long arm, on the other hand, would eventually be linked to a fair number of diseases, including forms of Lou Gehrig's, leukemia, epilepsy—and others.

It was while we were walking up and down the short arm that I came across a journal article that projected a very alluring reason to be interested in chromosome 21. It reviewed evidence that by the time most Down's patients turned forty, their brains usually were burdened with the same disturbing scene of amyloid plaques and neurofibrillary tangles that Alzheimer's disease produced. This begged a fascinating question. Did Alzheimer's, like Down syndrome, possibly spring from a DNA error on chromosome 21? Gusella's lab had the resources to find out. We had the chromosome 21 map and its probes, and we even had access to the cell lines of an early-onset Alzheimer family from Canada that had been collected by Ronald Polinsky and Linda Nee at the National Institutes of Health. Nee had sent the family to Gusella for mutation analysis shortly after Huntington's chromosome had been identified. Its ancestors originally from Northumberland County, England, the family represented one of the largest DNA troves of a multigenerational Alzheimer kinship so far gathered by researchers. At that point, very few family DNAs associated with any inherited disease had been amassed. Until the Huntington's breakthrough, there'd been no reason to collect them, no ready access into their faulty DNA.

If it hadn't been for the Huntington venture's proof of concept that mutant genes could be directly found, I sincerely doubt I would have so eagerly given chase to a major disease like Alzheimer's. Little was known about the genetic mechanism at work in its early variety, but judging by its inheritance pattern, the fault lay with an autosomal-dominant mutation, the same as in Huntington's. Alzheimer's was hardly trendy to work on. Yet its very obscurity as well as its tremendous torment to society made it all the more interesting to me. The Huntington's capture traumatized me into believing, rightly or wrongly, that anything in science is possible, and that high-risk ventures can be much more satisfactory to the soul than safer ones.

I'd seen Jim Gusella go after his own disease gene and more than survive. Here was an important gene I could tackle. In taking on Huntington's, Gusella hadn't only put himself out on a limb, but had also chopped off the strong branch he'd been standing on. He'd forfeited a signed-and-sealed position to postdoc in Leroy Hood's lab at Caltech in Pasadena, an opportunity others would have killed for. Hood, a high-powered molecular biologist who was inventing machinery that could automatically sequence DNA, had a mammoth reputation, and after working in his lab, Gusella might have had his choice of excellent jobs.

Another risk-taker I greatly admired was Stanley Prusiner, a neurologist at the University of California, San Francisco. Prusiner, whom I would soon befriend, was taking considerable heat for having proposed in 1982 that a certain crafty protein could do something proteins weren't supposed to do—change into an infectious agent that could be transmitted among animals and people and replicate itself. He called these proteins "prions," for *pro*teinaceous *in*fectious particles. So strangely, a person can be born with a gene mutation that apparently creates these infectious prions. (Prusiner's brave leap beyond dogma would win him a Nobel Prize in 1997.) Different forms of prions were responsible, Prusiner alleged, for a group of brain diseases that under the microscope are truly horrible to behold, for they leave behind gaping holes in brain tissue. Included among these "spongiform" diseases are scrapie in sheep; "mad cow disease" (bovine spongiform encephalopathy) in cattle; and, in people, kuru, Creutzfeldt-Jakob disease, and Gerstmann-Straüssler-Scheinker disease—the latter three comparable to mad cow disease.

Prusiner had even proposed that a prion bore such a close resemblance to Alzheimer's amyloid that their mechanisms might be one and the same: that Alzheimer's, therefore, might be an infectious disease. For many a year thereafter—even once the chemical differences between Alzheimer's amyloid and the prion were established—the idea hung around that prions might occur in Alzheimer's plaques.

Gusella gave the go-ahead, and off I went, rather innocently, to track down a behemoth: possibly a mutant gene on chromosome 21 that was responsible for Alzheimer's. One by one, by doing round after round of Southern blots, I ran each chromosome 21 probe against the DNA of the Canadian Alzheimer family to see if any variation markers consistently cosegregated with those who had the disease. The model of the Huntington's search preoccupied me; I expected that just as the Iowan family had for Huntington's, the Canadians' DNA would start giving us signs of an Alzheimer flaw on chromosome 21. It might take a while, but in time we'd tease out the incriminating evidence. Once the gene was hooked, way far out loomed the incredible idea of fitting a treatment to a mutation or a flawed protein. Drugs that still had years to go before completion were being made in response to the "cholinergic hypothesis"—the observation that a deficiency of the brain chemical acetylcholine might figure prominently in the disease. But they shriveled in concept compared to treatments based on the disease's genetic origins.

Housman, Gusella, and Wexler meanwhile had recruited an international troop of researchers to help find which specific gene on chromosome 4 contained Huntington's defect. Gusella's landing of the gene's approximate vicinity had happened so quickly that many people were optimistic that the mutation would be identified within a year or two. Because of Gusella's first-stage victory, the press's revelry still hadn't abated. Soon into the future—in large part because of the Huntington's achievement—one disease gene after another would be located faster than a robin finds worms. Back then, however, simply linking a disease to a portion of a chromosome was so novel it was monumental.

Stoked by the press, a sensitive side issue came to claim almost as much attention as the Huntington's discovery itself. By analyzing the bloods of a Huntington's family for the G8 marker, a physician had the potential to tell which children had inherited the chromosome 4 flaw. When a British doctor affiliated with Oxford University requested the G8 probe from Gusella for clinical testing, Gusella returned a cordial but firm "no." The probe was being made available to research centers, but "it wasn't appropriate to send it out for presymptomatic diagnosis because we didn't yet know if it could accurately diagnose Huntington's in all at-risk families," maintains Gusella.

Not long after, a letter by a group of Oxford-based geneticists appeared in *Nature* and contritely grumbled about Gusella's withholding of the probe. Remembers Gusella, "I got ripped—especially by their suggestion that with every new technology one needs expect a few casualties." The Oxfordians' exact phrase in *Nature*: "In no field of effective medicine can techniques be applied without casualties." Replying in the same issue, Gusella wrote, "We feel a moral responsibility to prevent the premature clinical use of the G8 probe until the information derived from it can be judged accurately. A scientist cannot ignore the social consequences of his work, especially in medicine." To control the probe's widespread use, Mass General filed for a patent and established a committee to explore guidelines for genetic testing and counseling. Eventually G8 was made available to clinics where responsible testing programs were in place.

To this day Gusella is considered a hero for not bending to pressure, for setting the invaluable precedent of paying heed to the ethical and legal safeguards that necessarily go hand in hand with genetic testing. Gene scans carry an assortment of possible backlashes for those tested, not the least of which is job, insurance, and societal discrimination. But in an era when scientists attract criticism for not being ethically and morally responsible for their discoveries, Gusella's actions sent the message that what scientists do shouldn't be just science, but science tethered to the needs of society.

By the summer of 1984, caught up in the search for Alzheimer's mutant gene, I'd finished running all the chromosome 21 map markers through the Canadian kinship, and Peggy Wallace, who previously had computed the Huntington's linkage, began analyzing the lod scores. Like an expectant father of a newborn, I joined her in Mike Conneally's lab at Indiana University to wait out the results. Computer programs in those dark ages were sluggish, and for days on end in the midst of the doldrums of a hot Indianapolis summer, she and I played contentious games of Scrabble to pass the time. In 1990, Wallace would receive recognition for assisting Francis Collins at the University of Michigan in nailing down the gene defect connected to neurofibromatosis, a disorder that causes largely benign internal and external tumors. Her prowess as a first-rate scientist carried over to the Scrabble board. Never once did she let a triple-score opening slide by.

Finally the numbers were all in. The answer was clearly, disappointingly: no. Chromosome 21 showed no signs of an Alzheimer-related defect, at least not in the Canadians. Although let down, I wasn't altogether surprised. Nothing incriminating had popped out from an earlier inspection of the raw data.

Despite this dead end, it was hard to shake the notion that an Alzheimer flaw sat on chromosome 21. The pileup of amyloid in the brains of Down syndrome people seemed too blatant a clue to abandon. Testing another Alzheimer family seemed obligatory. So in the summer of '84, with the help of several others in Gusella's lab, I began analyzing chromosome 21 in a second early-onset kinship recently sent to Gusella—an Italian family originally from the province of Calabria. Its collection had been largely undertaken by Robert Feldman, chief of neurology at Boston University School of Medicine, and Jean-François Foncin, head of neuropathology at La Salpêtrière Hospital in Paris.

Since first reporting on the family in 1963, Feldman had grown close to many of its members. He'd stood by helplessly as one after another came down with symptoms by age thirty-five, which is early even for early-onset Alzheimer's, and was laid to rest—as happened in this family—by forty-five. He was only too glad to see the family's cell lines passed over to Gusella. "Gusella was the only game in town in terms of being able to do molecular genetics for finding disease genes," Feldman recounts. "The sad irony" about those stricken in this family is that "they've all been super-bright, real achievers," he notes. This has raised speculation that their gene flaw might possibly sit close to, hence is coinherited with, a gene that influences intelligence. Today, the family's troubled past is known to extend back as many as ten generations, its descendants largely residing in Italy, France, and the United States.

One day after I'd started analyzing this family, the most interesting news swept in out of the blue. A small team on the West Coast had succeeded in determining the composition of Alzheimer's notoriously insoluble amyloid protein. The University of California team, headed by a pathologist by the name of George Glenner, had managed to dislodge amyloid from the brain's blood vessels. Right away I got hold of Glenner's report and was impressed by the strategies he and Cai'ne Wong had employed. Of even greater interest, Glenner and Wong confirmed in a second report that their retrieved *beta protein* was virtually identical to the amyloid protein that littered the cerebral hemispheres of individuals with Down syndrome. Furthermore, Glenner made a prediction that was music to my ears: "[This] suggests that the genetic defect in Alzheimer's disease is localized on chromosome 21."

I shared Glenner's reports with Jim Gusella, and we agreed that Glenner and Wong's retrieved beta peptide was an irresistible morsel. Here, through genetic-linkage analysis, we'd been scanning chromosome 21's entirety for an Alzheimer mutation, but now, by using the sequence of their short fragment, we could work backwards and go straight to its corresponding gene. My confidence came roaring back. Quite possibly we had simply skipped over a defective gene on chromosome 21 in the Canadians; quite possibly it remained true that one defective, defiling gene created a protein that not only clogged the brain with amyloid but initiated everything that went wrong in a disease I wouldn't have wished on my worst enemy.

After Julia gave birth to a tenth child, something changed. She was quieter, less cheerful, even despondent sometimes. Had childbearing just plain worn her out? Or was she perhaps suffering from postpartum depression? Her sisters, husband, and older children waited for her to "snap out of it," and indeed on some days she seemed her upbeat tireless self again, whizzing from errand to errand and baking up a storm. But even on her better days she kept forgetting small things. On numerous occasions she left the oven on, or lost the car keys, or confused days of the week. Clothes stayed out on the line long after they had dried. One afternoon she forgot to take the two youngest to a birthday party. Several times it slipped her mind that her husband said he'd be home for lunch. Her family teased her. "Boy, Mom, it must be early menopause!" "Boy, Mom, you'd better start tying a string around your finger!" Julia's laugh would bubble up. Really, she didn't know what had gotten hold of her.

— *four* —

Gone Fishing

Everything can be found at sea
According to the spirit of your Quest.
—Joseph Conrad

Word that George Glenner had gained a chemical foothold in the fibrillar brain deposits of Alzheimer's disease didn't take long to trickle into many other labs clear around the globe. Jim Gusella and I were unaware of it at the time, but quickly and quietly the ranks were forming to use Glenner's amyloid-beta peptide to backtrack to its corresponding gene—perchance, if mutated, the molecular answer to early-onset Alzheimer's inheritance. The ultimate vision was that if an errant gene did incite Alzheimer's, a drug that targeted its protein might derail the disease.

Over a dozen groups would take on the challenge. Following Glenner's example, bets were on that Alzheimer's was an amyloidosis of the brain. Not that a shred of hard evidence implicated amyloid in Alzheimer's, but since deleterious proteins connected to other genetic disorders were leading researchers to mutant genes, the seemingly abnormal protein fragment isolated by Glenner and Wong was worth the wager. Glenner also set his sights on using his peptide to lock onto its originating gene. Lacking the resources, he had turned the job over to a colleague in Japan, who, as several of the rest of us would do, began the preliminaries of fishing the gene out of the human gene pool. Meanwhile, Glenner got busy investigating whether the little extracted brain peptide could serve to diagnose Alzheimer's.

As for my own desire to pursue the amyloid gene, a period of personal transitions interceded. During the summer of 1984 I had taken two quantum leaps—I'd gotten married to Janet and I'd also quit the band. Playing and composing music would continue to claim me in spare moments, but I'd grown tired of the squabbles that kept erupting between our band troop and the various agents and club owners we depended upon. And operating on too little sleep was taking its toll. Besides, graduate school couldn't be postponed any longer. I'd been accepted into Yale's human genetics program, but Jim Gusella persuaded me that instead of more genetics, what I really needed was a thorough grounding in the neurosciences—an integrated spectrum of everything related to the nervous system. So withdrawing from Yale, I applied to Harvard's neurobiology doctoral program, got accepted in the spring of 1985, and prepared to osmose back into student life that fall.

Major changes were under way in Gusella's lab as well. The next phase of the Huntington's search—identifying the actual gene on chromosome 4—necessitated compiling more Huntington's kinships from all over the world and merging more experts from different persuasions into the equation. Gusella was overseas a great deal, coordinating these efforts. No longer was the gene quest a closely knit, one-house affair, but a giant international octopus with arms extending into laboratories on several continents.

We were still using time-consuming linkage analysis to search the length of chromosome 21 for an Alzheimer abnormality. Juggling several lines of research and anticipating that I might be leaving the lab for student life, I reluctantly turned over my tests of the Italian Alzheimer family's DNA to Peter St. George–Hyslop. A newcomer in Gusella's lab, Hyslop, in his early thirties, had grown up in England and possessed a sterling British accent and wit to match. He'd earned a medical degree from the University of Ottawa as well as a Ph.D. in neurology from the University of Toronto. Focusing increasingly on Alzheimer's disease, and narrowing that interest to genetics, he had written Gusella requesting training in DNA analysis methods, and Gusella was glad to oblige.

The first time Hyslop and I met, I was in the unit's coffee room entering genotypes—the two versions or alleles of a gene a person inherits, one from mom, one from dad—into a notebook, and Hyslop mistook me for a clerk. Soon enough, we trained him in running gels, blotting, and hybridizing, and before long he too was bent over his record book, scribbling down genotypes for the purpose of following variation markers through a family. Proof of his diligence at the bench, his lab coat got dirtier—and dirtier—and became the object of a certain amount of ribbing, which he cavalierly endured.

The plan of going after the amyloid-beta gene—perhaps the deep seat of Alzheimer's—was never far from my mind, and once I started graduate school in September 1985, I began looking for a way to fit the project into my curriculum. At Harvard it felt like a natural progression to at last be moving from small to big; from the underworld of genes and proteins I'd been immersed in to the structures and systems they give rise to. I'd decided that the molecular aspects of disease would nonetheless remain my focus, even though—and it didn't take me long to realize this—this cast me as a black sheep among the white rams of Harvard's neuroscience program. Although believed in by the faculty, molecular genetics was apt to be regarded as "cookbook" science. Some of our department's luminaries feared that students utilizing its trendy protocols, which were viewed as simple cut-and-paste DNA recipes as opposed to the more drawn-out procedures normally practiced by neurobiologists, would end up thinking that science was a snap. It was bad enough to be approaching neurobiology through the new window of genetics; worse yet to go beyond basic science and attempt to apply genetic minutiae to disease. What heresy! Much more to my advantage would have been to meticulously measure electrical currents in nerve cells and perform other procedures tied to the traditional bastion of neurobiology—neurophysiology. Several senior faculty had helped pioneer this field, which explores a nerve cell's ability to transport and send electrical signals. As would become increasingly obvious, the hardest part of my grad-school passage wouldn't be the science; it would be the adversity that developed between me and a few of the program's more vocal higher-ups.

One of the only other students in the neuroscience program concentrating on disease genetics was Anthony Monaco. Monaco, who got hooked on neuroscience during his undergraduate years at Princeton, was also of Italian descent. I soon noticed that some in the program didn't bother telling us apart. "Hey, Tony," certain faculty repeatedly hailed me as we passed in the fourth-floor hallway of Harvard Medical School's Building B, "how's the gene stuff going?" To others we were just those dark-haired Italian guys applying cookbook science to profound neurological problems.

The opportunity to isolate Alzheimer's amyloid gene at last arrived when, at the start of 1986, I began my first rotation under David Kurnit, a human geneticist at Children's Hospital. I'd requested Kurnit as my rotation adviser because of his expertise in chromosome 21 genetics and Down syndrome. He accommodated the plan to use Glenner's peptide to backtrack to its corresponding gene, even though the neuroscience program's director Edward Kravitz had his concerns, as did others. Perhaps neither amyloid nor any of the other proteins stuck in the plaques would lead to a mutant gene that commenced the disease. "Watch out," people advised me. "The amyloid associated with this disease might be just garbage."

But if I was deluded, at least I had company. Kurnit, seeing an obvious coupling, set me up with one of his postdoctoral fellows, Rachael Neve, a high-adrenaline, hardworking, frizzy-haired molecular biologist who, newly captivated by memory, learning, and other mysteries of the brain, also happened to be working on a game plan to track down the amyloid gene. As pragmatic as it might sound, it was an obvious marriage of convenience, for each of us had skills and resources the other lacked. By then it was clear that several other groups with far greater clout, including some prominent drug companies, were bearing down on the amyloid gene. As for Neve and me, "There was a good synergism between them," remembers Kurnit.

We weren't using variation markers and genetic-linkage analysis to narrow down the disease gene's location on chromosome 21. That approach—indoctrinated by the Huntington's project—was being used across town by Peter Hyslop at Mass General. Instead, Neve and I, employing the older method of gene snaring, were leaping on the back of a disease-associated protein and hoping to ride it directly to its potentially defective gene. From time to time I'd check in with Hyslop to learn whether he was seeing any DNA discrepancies in the Italian and Canadian Alzheimer families. Emotionally, it had been hard turning the linkage work over to him, given that it was an extension of my early navigation on chromosome 21. But it didn't help that Peter was becoming increasingly private about his progress.

As in certain other competing labs, Neve and I began employing chemical bait—oligonucleotide probes based on Glenner's found peptide—to try to fish out its corresponding gene from libraries of normal genes. Oligos are short stretches of DNA. Since fall, on and off again Paul Watkins at Integrated Genetics and I had been working to design and synthesize them. This amounted to trying to correctly guess the bases in the amyloid gene that encode its peptide's beginning and end sections. Researchers in the 1960s had elucidated how genes make proteins: a set of three consecutive bases in a gene—a codon—encodes each amino acid in a protein. This invaluable gene-to-protein translation represents the "genetic code." However, most amino acids result from not just one codon, but any one of two to four possible codons, which is why our work required endless guestimating, a wing and a prayer.

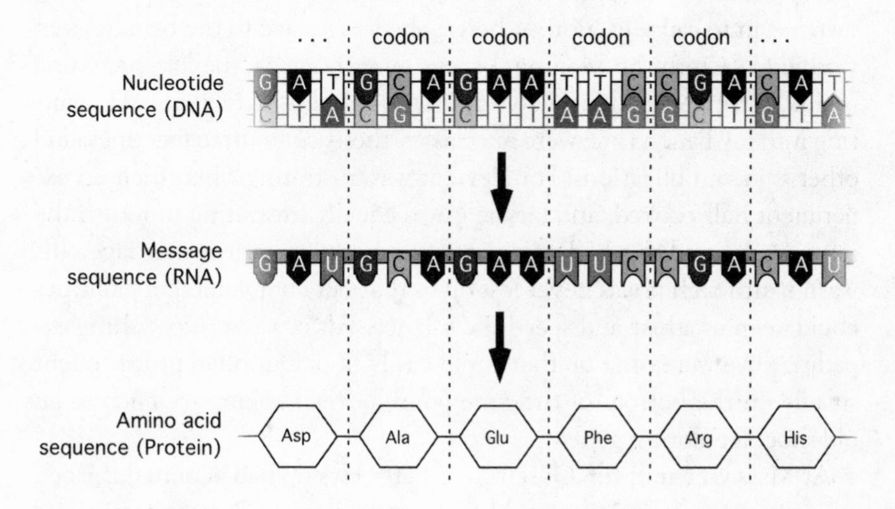

FIGURE 4.1 DNA to RNA to protein. Illustration: Robert D. Moir

While the oligos were my offering to the research, Rachael Neve's expertise gave us a pond, or library, of all the genes "expressed" or active in a certain organ to fish from, so that when we lowered our bait into this pond, we might lift out the gene we were after, or at least part of it. Neve was a maven at making these libraries. They essentially provided us with the protein-encoding regions of genes, without the intervening "junk" DNA. Her ability to extract high-quality RNA—the intermediary for a gene's conversion into a protein—was critical to our progress. Working with RNA takes real talent, because unlike two-stranded DNA, RNA is one-stranded, wimpy, and easily breaks down. If you so much as touch glassware, the oil left from your fingers can degrade RNA's fragile structure in seconds.

It wasn't long before Neve and I discovered that we shared a liking for jazz, new wave, and cutting-edge rock. One of our indispensable pieces of equipment was Neve's high-quality Bose Acoustic Wave boom box, and on any given day, at any given hour, the loud tones of the Cure, Keith Jarrett, Peter Gabriel, or the Smiths poured out the open door of our tiny, one-and-a-half bench lab on the fourth floor of Children's, filling the hallway and raising the blood pressure of other researchers: "Turn it down in there! What do you think this place is, a disco?"

It also didn't take long before differences in Neve's work habits and my own began to cause friction. Whereas she was a slave to the bench, often working late into the evening, I spent fewer hours in the lab, preferring to mix in trips to the library to stay on top of published reports. Also cutting into my bench time were my classes, thesis committee meetings, and other student obligations. For Neve, it was frustrating when I left an experiment half-cooked, and for me it was equally frustrating to return the next day to find she had completed my work. A perfect marriage ours wasn't, although it was never lost on us that our complementary abilities could keep us afloat and speed the science. For it was a risky fishing expedition we were on, one that could easily flop. Our oligo probes might sit idle on the bottom of the gene pond, never hooking so much as an old shoe, let alone a gene.

At Mass General, Jim Gusella and Peter Hyslop had acquired the cell lines of two more early-onset Alzheimer families—a German family and an extensive Russian family. Similar to his analyses of the Italians and Canadians, Hyslop had started employing linkage strategies to try to detect a defect along chromosome 21 in these new kinships. The anticipation grew that either by Neve's and my straight-line protein-to-gene approach, or by Hyslop's linkage route, or both, something major might be pulled into view.

Concurrently, Harvard's neuroscience program was beginning to have the desired effect. My prior molecular orientation was being stretched outward. I was learning my way around the nervous systems of various organisms—the simpler nerve pathways of invertebrates, including drosophila, geneticists' favorite fruit fly; grasshoppers; and lobsters; then onward to vertebrates: mice, rats, and the vastly intricate neural circuitry of humans. The more I grasped about the interconnectedness between genes and proteins and brain cells and the brain's grosser structures, the more I tried to plug every new piece of information into a scenario of how an altered amyloid gene might relate to Alzheimer's. A new idea would strike me, I'd run to the library, photocopy dozens of pages from various research reports, speed back to the lab, and try out my *Hypothesis of the Day* on Neve. This exercise actually became an effective means for learning more about the brain than I might have otherwise. One day it occurred to me that maybe amyloid's gluey protein served a purpose by facilitating connections between neurons, but in excess it gummed up the works. Another day it hit me that maybe amyloid reinforced essential calcium currents in nerve cells, but that, again, too much of it was a bad thing. In retrospect, both hunches amounted to what researchers refer to as "shameless speculation." Nonetheless, Neve, always accommodating, would provide thoughtful support or rebuttal.

On other days, the idea of disentangling the components of Alzheimer's origins seemed overwhelmingly complicated. As my graduate studies were making ever so clear, the human brain was so complex, it made computers of the day seem like child's play. And that was a healthy brain.

Our brain, I learned, is a monument to our species' ability to *move* and cognitively deal with the consequences. Quite literally, it's an electrical, chemical universe wherein the wizardry of thinking and remembering, consciously and unconsciously, depends on as many connecting neurons, or nerve cells, as there are stars in the Milky Way. Or maybe even more. Some say the human brain holds as many as 100 billion neurons. Others say twice that. It's a hard number to count. The brain's glial cells (from the Greek for "glue"), which keep neurons nourished and lodged in place, are many times more numerous. Maximizing the number of neurons, the cerebral cortex—the thin outer mantle of gray matter that's crowded with neuronal cell bodies (hence "gray")—is replete with folds. As humans continue to evolve, more than likely these folds will keep increasing, making room for all the more neurons and furthering our intellectual capacity.

Attesting to the human brain's extraordinary complexity, as many as 30 to 50 percent of our roughly 100,000 genes are active in the brain.

All the proteins these genes encode, therefore, are dedicated to its development and performance. To digest and process food requires a fair number of genes and related proteins. But imagine the numbers necessary for reading and thinking through a book like *Moby Dick*—for recognizing and understanding its multitudinous symbols and associating them with the world around you. Our brain's ability to think rationally and not simply react, as do many species, is made possible by a quadrillion connections supported by trillions of nerve fibers. For the human brain is nothing if not connections. To receive a message, a neuron grows an abundance of sinewy branching arms, or dendrites. For sending impulses, it utilizes an often longer, thinner protruding arm—its axon. Both armlike extensions, or neurites, allow a neuron to relay messages with many thousands of other neurons.

As my studies showed me, Alzheimer's attack on the brain's intricate wiring is anything but subtle. Certain brain regions can lose 30 to 50 percent of their neurons. Millions of cell bodies shrivel, die, and are reabsorbed into the extracellular fluid, simply vanishing. Degenerating dendrites and axons of living and often ailing neurons, some attached to a neuron's body, others detached, get caught up in the slurry of forming amyloid "neuritic" plaques. In a matter of years it would be definitively shown that in Alzheimer's the essential loss to the brain and body isn't the neuron and its tentaclelike arms, but the disappearance of the synapse—untold numbers of synapses. This is the juncture where an impulse, or message, reaches the end of a neuron's extended arm and chemically vaults across a sliver of space called the synaptic cleft and into the outstretched arm of another neuron. In the words of Ramon y Cajal, who many regard as the "Father of Neuroscience," a "protoplasmic kiss."

In the central nervous system, which encompasses the brain and spinal cord, often thousands of synapses must occur for one message to reach its final destination. The more synapses along a neural pathway are utilized, the more memories appear to be reinforced and stored long term in the cerebral cortex, most researchers believe. The precise mechanisms of memory and its storage and retrieval are far from understood, yet it's not hard to appreciate that when synapses are grossly lost in brain regions central to memory and cognition, so is a person.

As I studied this neuronal havoc, I kept asking myself the same question that had been asked intermittently ever since Alois Alzheimer's day. Were, in fact, amyloid plaques, as they collected between neurons and atop the fibers that crosslink them, responsible for disconnecting or directly killing neurons? Even though Glenner and others suspected as much, no one had yet shown that too much amyloid was toxic or detrimental to brain cells.

Or could it possibly be that the flame-shaped neurofibrillary tangles inside cells were the disease's chief culprit? Before the 1950s, the tangles often were assumed to be residual riffraff from the disease. Then, midcentury, they began attracting more attention. Their formation inside neurons suggested they were more apt to do damage to neurons than the plaques located in the gelatinous space between neurons. But by the mid-1980s, the plaque was the lesion most investigators were inspecting, and not simply due to Glenner and Wong's extraction of the beta peptide from plaque cores. The tangle, it turned out, was even tougher to crack than insoluble amyloid—a surprising discovery made in 1981 by Dennis Selkoe, then at McLean Hospital. This made it extremely difficult to isolate and identify the tangles' subunit component, as Glenner had done for amyloid plaques. In fact, researchers before Selkoe who thought they had gotten hold of the tangle's base unit had only been fooling themselves. The tangles' core ingredient remained unknown.

Selkoe and various other Alzheimer researchers would be further dissuaded from pouring all their efforts into the tangles because several other brain disorders exhibited them as well. So it seemed less likely

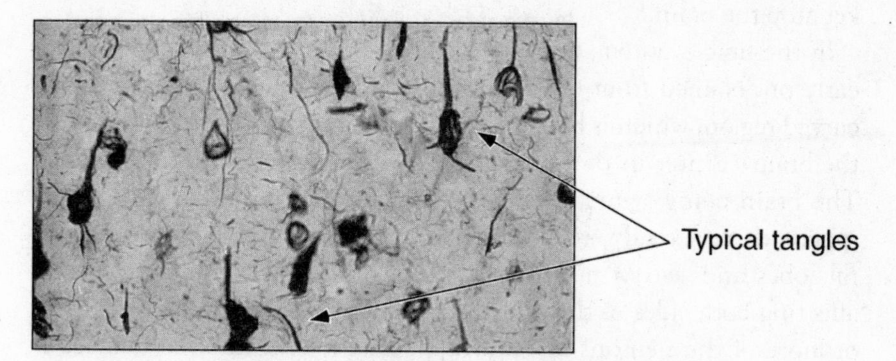

FIGURE 4.2 Photomicrograph of neurofibrillary tangles inside neurons.

that these knots drove Alzheimer's pathology. Pick's disease, Lou Gehrig's disease, Down syndrome, Guam Parkinson–dementia complex, dementia pugilistica—a dementia brought on by trauma to the head—and other ills had the tangle in common. Amyloid, by comparison, seemed a far more unique signature in Alzheimer's. Down syndrome was the only other disease that lodged such vast quantities of amyloid in *both* the brain's blood vessels and between cortical neurons. (A handful of other diseases were known to accumulate amyloid mainly in the brain's blood vessels.) Moreover, it looked as though tangles developed later in Alzheimer's, after something else precipitated neuronal cell death. "Amyloid, on the other hand, appeared possibly to be an earlier event," points out Selkoe, and therefore might be the precipitator. When, in 1984, George Glenner gave science the gift of the beta peptide isolated from plaques, even more researchers switched their emphasis to amyloid.

I'd read that certain brain regions could be crowded by plaques and tangles, while other regions that lay right next door could remain relatively unscathed. But as I sectioned my way through diseased brains, it was a whole other thing to witness this curiosity. It brought to mind a street a tornado passes down, the houses on one side untouched, the houses across the way in ruins. But unlike a tornado, the disease follows a more or less predictable path, almost as if it's attracted to a particular molecular substrate. Especially vulnerable is the gray matter, or neuron-rich tissue, in two brain regions: the limbic system, a group of structures tied to memory and emotional behavior that extends from deep in the brain outward; and broader regions of the cerebral cortex, the thin blanket atop the brain.

In the limbic system, the hippocampus is severely affected, and quite early on. Named from the Latin for "seahorse" due to its shape, this curved region, which is barely one-and-a-half inches in length, lies near the brain's center in the temporal lobe—straight in from your temple. The brain being symmetrically divided into a left and a right hemisphere, there actually are two hippocampi, just as there are two temporal lobes and two of most every region under the cranium. Usually affecting both sides of the brain, Alzheimer's often destroys 50 percent or more of the neurons in both hippocampi. This internal damage is outwardly mirrored by a victim's worsening symptoms of forgetfulness, for a working hippocampus is essential for capturing and converting new information that is stored long term in the cortex. In 1953, the famous case of H.M. revealed the hippocampus's great service to memory. When H.M., an epileptic, had two-thirds of each hippocampus removed, his seizures stopped, but at great cost. While his formerly preserved memories remained intact, he no longer could retain new facts or recall recent events. Since a working hippocampus also supports spatial orientation, H.M.'s sense of location and direction suffered as well.

The disease's mysterious selectivity is very apparent in the hippocampus. Certain of its layers can be riddled with plaques and tangles, while neighboring others are more or less spared.

Two even smaller limbic regions are as severely affected: the entorhinal cortex and the amygdala. A gateway to the hippocampus, the entorhinal cortex may be where Alzheimer's tangles first appear. As its tissue degrades, its memory signals are alienated early on from the hippocampus, never reaching the cerebral cortex. Alzheimer's advance into the amygdala, a region directly below the hippocampus, happens later in the disease process. The amygdala, only about the size of a garden snail, is thought to attach emotion—be it fear, anger, or passion—to memorable events. Alzheimer's completely ravages the amygdala, which may explain a patient's unnatural emotional responses and personality changes. But then again, so might the disease's severe damage to the entire limbic system, which includes many more regions than those mentioned.

To illustrate Alzheimer's progression through various limbic regions to his Harvard medical students, MGH pathologist Jean-Paul Vonsattel draws from a suitor's conquest. "Early on, Alzheimer's overtakes the hippocampus—the horse, the symbol of power," Vonsattel relates. "From there it moves to the fimbria and the fornix." Fornication. Thoughts thereof enter the suitor's mind. "From the fornix it travels to the mammillary body in the brain." The breast. And so it's on to foreplay. "From the mammillary body it travels to the thalamus, which in Latin means sleeping room, where the anterior nucleus and dorsal media are prone to degeneration." The rest of the thalamus is relatively untouched. "Then it proceeds to the cingulate gyrus. Cingulate means belt, you know." The belt loosened. "And it's on to the amygdala—amo, amas, amat." Love is made. Not that Vonsattel takes the disease lightly. Having peered via autopsy into over 8,000 human brains, he knows all too well the severity of Alzheimer's assault. "All main parts of the limbic system are terribly, markedly involved," he notes.

From limbic terrain, Alzheimer's pathology travels into the cerebral cortex. Even though this outer-brain region is less than one-third of an inch thick, the expansiveness of its network of neurons is Nature's gift to humans. Here, the temporal lobe (memory, learning, emotion, hearing), the inner part of which contains certain limbic areas already discussed, is especially overtaken by plaques and tangles. Lesions also occur in the frontal lobe in the front of the brain (judgment, reasoning, planning), and to a lesser extent in the parietal lobe (touch, pain, temperature), which lies above the temporal lobe. In the disease's cryptic way, it leaves the hindmost lobe—the occipital, where visual information is processed—relatively undisturbed.

Although certain zones of the cortex are responsible for specific processes, researchers currently are realizing that the cortex's response to

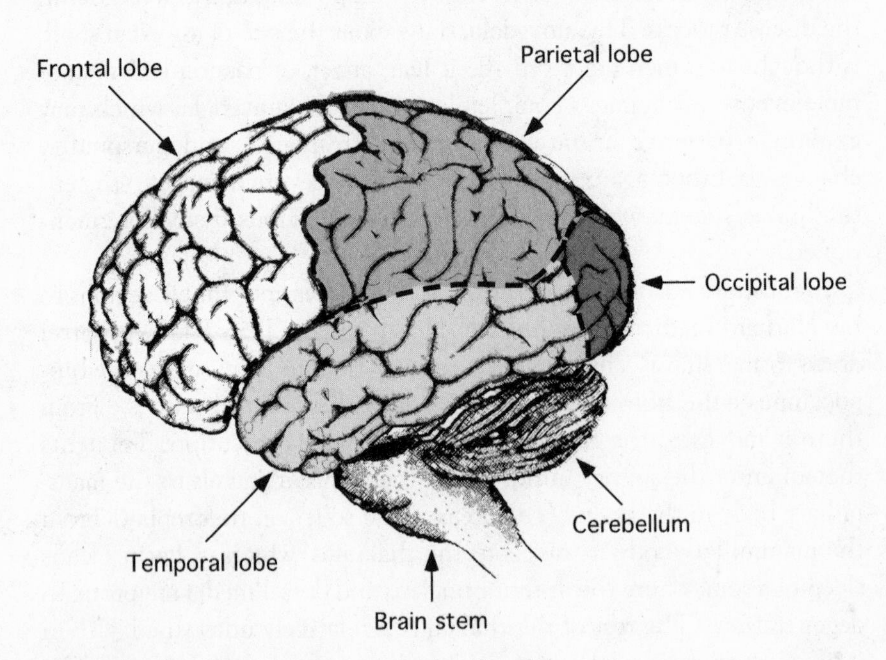

FIGURE 4.3 The human cortex and its four lobes.
Illustration: Robert D. Moir

stimuli often conjoins neural pathways from different regions. While a response supported by memory may be the sum of many parts, generally the left cortical hemisphere appears central to language and planning abilities; the right to spatial adeptness, notably creativity and intuition. Because of the left brain's more practical abilities, neuroscientists are apt to advise, "Don't leave home without your left brain!" Since Alzheimer's infiltrates both sides of the brain, the cognitive cargo in both are usually lost, unlike in Pick's disease, which can be asymmetrical and do more damage to the left side.

The brain tries its utmost to bounce back from this death march. When neurons die, others nearby attempt to compensate by sprouting new connections to regain lost synapses and maintain the integrity of the neural network. I was astonished by how this lush regeneration in the Alzheimer brain could resemble the fervent sprouting of fibers seen in a young child's developing brain. Years ago, when our family vacationed in our beach house in Narragansett, Rhode Island, the old man living next door kept getting lost in the woods out back. My parents told me and my twin sister it was because he was going through his second childhood, which at the time confused the heck out of us. Now as I witnessed how the growth of new connections in a wasted brain resembled that of a very young brain, a "second childhood" seemed a particularly apt description for our neighbor's probable Alzheimer's.

One day in my Neurobiology of Disease class, a case study wrenchingly drove home Alzheimer's very worst aspect—its evisceration of a person's former self. The man led into the room and seated in a chair was in his early fifties. He'd been a buyer of apparel for a large department store, and the disease had pushed far enough into his cortex that he'd forgotten the vocabulary he had once used with regularity—words for the clothes he was wearing; simple words like "belt" and "shirt." It was a chilling display, and I'd like to think he was only minimally aware of why he was in our midst. When you meet an Alzheimer patient or think about what the disease destroys, it's impossible not to feel compassion—and a dull terror due to imagining where you'd be if your own world fell apart in similar fashion. I find in my work today that compassion, however, isn't always advantageous to a researcher. The emotions raised tend not to foster the objectivity of mind needed to effectively lead you toward a therapy that will help patients. It's a gut feeling I have, but the wits and energy you need for research seem better sharpened by passion for solving the puzzle as opposed to sympathy for the condition. In the same sense, a good martial arts fighter never wants to attack out of anger.

What makes us fear Alzheimer's so? Erase memory, we instinctively realize, and a person's very humanness is cancelled out. "Our reality is basically our memory," surmises James McGaugh, a prominent neuroscientist at the University of California, Irvine, who specializes in learning and memory. "We believe our existence is so continuous, so seamless, but really it's only a half a second long. Everything that happened a second before is memory, and the next second—what we believe will happen—is also memory. I regard memory as the bridge between the past and present and future." Due to memory, then, we live in a delusional sort of way, projecting a temporal and spatial reality that is far, far more extensive than the dot of time and space we pass through at any given moment. Take away that delusional continuity between past and future, as well as time and space, and you begin to gain a sense of an Alzheimer patient's terribly limited world. As John Bayley wrote in *Elegy for Iris* about his wife, novelist Iris Murdoch, before she died of Alzheimer's, "she feels a vacancy that frightens her by its lack of dimension."

So many terms have sprung up for the various forms of memory that humans seem to depend on, some researchers unabashedly point out that it reflects how little is understood about memory. "There's a saying that one scientist would prefer to use another scientist's toothbrush rather than his terminology, and I think that characterizes this field," says Mark Baxter, a behavioral neuroscientist at Harvard. "What all these terms are especially trying to capture is the difference between conscious recollection and those memories whose learned origins we don't remember." Memory you consciously pull out of a hat, versus that which bubbles up without conscious effort. Calling up someone's name versus the automatic execution of the backstroke.

In Alzheimer's, consciously accessed memory is what so noticeably degrades. The inability to retain new input, leading to a chronic stumbling of short-term memory, usually is an early symptom. Yet as the disease advances into the cortex, all avenues tied to memory progressively give way, even those most unconscious and innate.

Having sailed into an investigation of Alzheimer's, I felt at this point like a Lilliputian up against the giant multiheaded Hydra. The disorder showed so many different faces. Dying nerve cells. Snuffed synapses. Severed pathways. Fallen brain regions. Impaired mental and corresponding physical abilities. Somehow these were all part of a string of disasters rooted in genetic abnormalities and/or environmental stressors. That its inheritance might all boil down to one lone gene mutation, as seen in Huntington's, seemed rather extraordinary.

Researchers were concluding that Alzheimer's rare early-occurring variety indeed was usually inherited. Studies from the late '50s of Scandinavian families struck down by this form had originally lent weight to this perception, with numerous other early-onset families—just like the four undergoing analysis at Mass General—rapidly being recognized worldwide. The genetic problem in these families bore all the signs of a generation-to-generation pattern indicative of a dominant mutation. Housing such a mutation, a faulty gene gets passed down by one parent, overriding a normal counterpart gene from the other parent and conferring the disease, on average, to half their children.

As for Alzheimer's extremely common late variety, my classmates and I were led to believe that it instead was *sporadic*—not genetic, but acquired over a lifetime, possibly because of one or another environmental insult that was toxic to the aging brain. But as I'd discover in time, detailed studies were telling a very different tale about late-onset Alzheimer's. For there were signs that it too could be genetic; that in some families a person who had an elderly parent with the disease stood a significantly greater chance of also getting Alzheimer's in later life. Whatever the transmission mechanism at work, it lacked a dominant mutation's strident regularity. It didn't necessarily hit as many offspring, nor even every generation, and seemed a more complex stew of far subtler genetic factors than its early-onset counterpart.

Many in the scientific community resisted the idea that late Alzheimer's could be genetic, however. They held firm to the belief that had prevailed since the disease had been forced out of hiding in the elderly—that environmental and/or aging factors were behind it. A small minority of others in the mid-'80s suspected Alzheimer's arose due to an infectious agent—a virus, a bacterium, or, more exotically, the prion that transmits Creutzfeldt-Jakob and other spongiform brain diseases. A possible infectious avenue was bound to be explored. The viruses behind polio and the recently exploding AIDS virus were models of human destructors.

By May 1986, just as my classes were winding down for the semester, Rachael Neve and I had in hand two oligo probes fashioned by Paul Watkins that looked to be working. They were binding tightly with various pieces in Neve's state-of-the-art libraries of expressed brain and liver genes. Simultaneously, subjecting the oligos to a novel technique we called the "genomic window," we were gaining evidence that Glenner's brain fragment—beta peptide—most likely did arise from a gene on chromosome 21, just as Glenner had predicted. But this evidence was so circumstantial we didn't dare get too excited. Meanwhile, voices around me didn't let me forget the hidden perils of going after the amyloid gene: "You know, it could turn out to be a wild goose chase."

A significant genetics achievement came to fruition in the Children's Hospital lab of geneticist Louis Kunkel in the spring of '86, and although the methods used were different from Neve's and mine, it provided inspiration. Kunkel and coworkers managed to isolate a piece of a gene on the X chromosome that contained a DNA deletion responsible for an inherited form of the fatal muscle-wasting disease Duchenne muscular dystrophy. The report of their priceless achievement would follow in October in *Nature*. Neve had supplied libraries of muscle DNA for the finding, and Tony Monaco had served as Kunkel's right-hand man throughout, gathering and analyzing the family DNAs and fishing out the gene. It was immensely satisfying to see Monaco help score a major win for human genetics, which all but established his career. Currently the director of the Wellcome Trust Centre for Human Genetics at the University of Oxford, he and his lab are combing the genome for genes connected to autism, dyslexia, and other related disorders.

The Duchenne finding hadn't fully depended on genetic linkage. Nor, as seen, was Neve's and my protein-to-gene hunt following that route. After the Huntington's 1983 breakthrough, the next major discovery to utilize variation markers and linkage analysis to track a gene had occurred two years later—in 1985—when a British team linked polycystic kidney disease to chromosome 16. "The Huntington's breakthrough came so early and so unexpectedly, the following two-year lag was a measure of how long it took people to collect the DNAs of disease families so the new gene-finding approach could be used," Jim Gusella points out.

Motivated and armed with the necessary resources for netting the Alzheimer gene, Neve and I nonetheless were having a frustrating time of it. Casting our oligos into Neve's pool of brain genes, we kept pulling out duds—random genes that didn't encode for Glenner's amyloid peptide. All the while the sound of hoofbeats grew louder—rumors that other teams were close to isolating the gene. We were constantly aware that at any moment we might find ourselves the losers. "You can be 99 percent of the way there and someone comes along and scoops you," Neve observes. "Because of that, gene hunts are very scary."

As spring turned to summer, I began noticing that every time the Red Sox won, our experiments went well; when they didn't, our experiments were apt to flounder. Expectantly, we took to tacking up the headlines following each game as a way of gauging our bench progress, the wonder of it being that the Sox kept getting stronger and stronger. "Red Sox Win Fifth in Row." Increasingly we were fastening onto promising DNAs, one of which might correspond to the amyloid protein's gene. "Clemens Pulls Up His Red Sox Again." Screening tests began to confirm we were on the right track. Then came good and bad news. "Red Sox Cool Tigers;" "Indians Overpower Red Sox." Tests showed that certain of our isolated genes might be from chromosome 21, yet the results implicated other chromosomes as well. But then the Sox really turned it on. "Red Sox Win Ninth in Row." As the red-hot home team batted its way toward a World Series playoff, in the lab we had what I believed just might be a piece of the amyloid gene. But the coding stretches we'd isolated from it appeared to be entangled with coding stretches from other extraneous genes, and due to this complication Paul Watkins, who was sequencing our fished-out clones, hadn't yet confirmed that it corresponded with Glenner's beta peptide.

In October, when the Red Sox faced off against the Mets in the World Series, superstitious sort that I am, there was suddenly a reason to doubt our recent progress. On Saturday, October 25, during the now-renowned sixth playoff game, the Sox's dream of winning their first World Series in sixty-eight years came crashing down. In the tenth inning, a slow grounder, which should have been an easy out for the Sox, rolled right through the legs of first baseman Bill Buckner, allowing in the winning run for New York and depriving the Sox of victory. "It bounced and bounced and then it didn't bounce; it just skipped," Buckner tried to explain to reporters—the man whose knees were so bad, anyone else would have been on crutches, his critics chided. I couldn't help but agitate over what the Red Sox's demise portended for the fished-out gene.

Two weeks later, off Neve and I trotted to the sixteenth annual meeting of the Society for Neuroscience, science's equivalent of a Convergence of the Clans of the Cavemen. Memorably, it was held in Washington, D.C., that year. Like Stone Age tribes of old, who journeyed for weeks to come together in one large place where they would trade tools and pelts, sightings of woolly mammoths, and rumors of other tribesmen, every fall thousands of scientists from all over the country and beyond made the Neuroscience pilgrimage to whatever port of call. (More than ever, they still do. In 1999, the conference attracted over 24,000 participants—up from 10,471 in 1986.) For month upon month they'd honed their data in anticipation of the meeting's hurling contest of new findings. Many a lab chief proudly brought along their youngest, brawniest, most skilled tribesmen, and when the competition began, in observance of the traditional rite of passage, chiefs often decorously stepped aside, allowing these promising postdocs and junior faculty the opportunity to present hard data and show their manhood, or in rarer instances their womanhood. For at the time, women were few and far between at these affairs, indicative of how few rose to chieftain status or were admitted into a lab's circle of fire. (An informal survey reported that in 1982 women made up some 20 percent of the attendance; by 1995, 30 percent.)

The best pearls were saved for Neuroscience. It was where labs sprang their most dazzling evidence and secreted surprises. The more earth-shaking and irrefutable the hunted-down data, the more respect and bragging rights a lab garnered, especially if it had submitted an abstract of the findings months in advance, proving it had speared its catch before anyone else. Just as reputations got made at Neuroscience, they also could be broken. If no one believed the material presented, it could undermine a lab's stature and the credibility of its hunters, dimming the prospects for future grants and publications. This annual conclave also was where younger tribesmen, the postdocs, sometimes met their future and got married off to other camps, which lessened the chances of inbreeding and ensured that the field remained fit and productive.

Molecular geneticists being a fairly new breed, few were invited to present papers at Neuroscience in those days. Those that did mostly gave descriptive reports on which genes got switched on in various parts of the brain for the making of proteins. I'd brought along autorads displaying the amyloid gene candidates, hoping to find some spare moments to continue my visual assessment. Among them was the clone I was willing to bet was the gene, although it still needed confirmatory sequencing by Paul Watkins. Everywhere I went I clutched my briefcase under my arm as if shielding the data from what I feared most: someone else delivering proof that they'd latched onto the gene. But no talks on the research were scheduled. Just maybe we would beat out the competition.

Midway through the conference, the chairman of the Alzheimer's disease research forum announced that someone by the name of Dmitry Goldgaber from the National Institutes of Health was going to make an unscheduled presentation of a late-breaking development. The session drew such a crowd that, arriving a hair late, I barely found standing room. Goldgaber was introduced—he worked out of the lab of the esteemed virologist and Nobel laureate Carleton Gajdusek—and up on stage strode a hefty, grayish-haired fellow. His words resonating through a thick accent, he came straight to the point: His team had isolated a portion of Alzheimer's amyloid gene, it was on chromosome 21, and here was the DNA sequence as proof. The news stunned everyone in the room. During his brief talk, no one moved, coughed, or sneezed. Goldgaber gave more description, then jogged off stage, at which point the room broke into a loud buzz. No one could believe what had just taken place, least of all me. I was as excited by Goldgaber's words as anyone present, but at the same time disbelieving of how this unknown player had stolen the show and gotten first bragging rights in less than five minutes!

"It was an incredible scene. It was the first seminal event in the field that unified pathology with genetics," remembers Kevin Kinsella. A venture capitalist blessed with brains, tall good looks, and an open, gracious style, not to mention a Midas-touch for starting up biotech enterprises, Kinsella at the time had just founded Athena Neurosciences, a company in California's Bay Area uniquely devoted to pursuing drugs and diagnostics for Alzheimer's disease. Bowled over just like the rest of us by Goldgaber's disclosure, on his flight back to San Diego that afternoon, Kinsella decided to call the medical editor at the *San Diego Union* and share the scoop. "These were the early days of sky phones, and they were expensive and difficult to use. I got connected, but the guy on the other end showed no excitement. He didn't have a clue as to why the finding was so important. This said something about him, but also about the profound lack of appreciation at the time for what genetics could mean for the future of medicine."

Overly eager and still wet behind the ears at twenty-eight, I approached Goldgaber after his announcement—people were sticking to him like flies, all wanting to know more about his clone—and flat out shared my own team's news. We had data, albeit circumstantial, indicating we too had snared the amyloid gene, I told him. And by using the MGH's collected families and the chromosome 21 map, we'd soon test whether our candidate gene was the main culprit behind familial Alzheimer's, the inherited form. Goldgaber invited me over to his lab at NIH to discuss our work further, and together we headed for the subway.

Born in Latvia, Goldgaber had studied molecular biology and physics at Leningrad Polytechnic Institute before arriving in the United States in 1980 and joining the NIH lab of Carleton Gajdusek. Gajdusek, an intense, highly loquacious graduate of Harvard Medical School, had attained almost mystical stature for determining that kuru, a disease that eats away at brain tissue, was transmitted by an infectious agent and arose, as observed in a tribe in New Guinea, from the ritual of consuming the brains of the dead. For his inroads into kuru and other brain disorders, Gajdusek was awarded a Nobel Prize in 1976. Stanley Prusiner would later cite kuru among the spongiform diseases he attributed to prions—uniquely infectious proteins.

Within Gajdusek's lab, Goldgaber introduced me to a human geneticist on his team—Wesley McBride—and he and I compared autorads from our respective labs. The DNA fragments depicted on both labs' films indicated the very same gene! Very remarkably, for all we knew we could be staring at the cause of inherited Alzheimer's. Although at that point no mutation had been spotted on the amyloid gene, imaginably one existed that made for an errant protein that so demonically aggregated in the brain. Had they wanted to investigate whether the gene had a mutation, the Gadjusek-Goldgaber team had no immediate avenue, since they lacked the necessary DNAs of a family collection. Neve and I at Children's Hospital, meanwhile, had access to the Mass General families due to our collaboration with Gusella's lab.

Going to the trouble of taking the tube containing his team's sequence of the amyloid gene out of the fridge, Goldgaber made a suggestion. Why didn't I take his clone back to Boston and, using the MGH's Alzheimer families' cell lines, together we would do the necessary tests to see if a particular version of the gene coinherited with those who had the disease. If one did, it must hold a mutation. A collaboration would have been nice, but it didn't make a whole lot of sense to me since our Boston group already had its own clone. Later, when we went outside to where a taxi was waiting to take me to the airport, Goldgaber still had the tube in his hand. No obligation, he said, as he handed it to me. But I very much felt trapped into an obligation, and the whole way home came close to tossing the tube into the nearest trash bin. I sensed that in accepting it I'd made a huge political blunder.

In frequent phone conversations with Goldgaber in the weeks following, I kept explaining that since we already had the gene, our team didn't feel a collaboration was feasible. Paul Watkins had completed the sequencing of our best candidate—which was 1,100 bases long—and chemically confirmed it was a piece of the gene that encoded the amyloid protein. Some months went by. Then, around the time both the NIH account and our separate account of the gene finding were being published, I began hearing something disturbing from other colleagues. The NIH team reportedly was spreading the word that they had given our Boston team their amyloid gene probe—the grave implication to the field being that we hadn't found the gene on our own. To put it mildly, I was thrown for a loop. In retrospect, I realize I was naive to think that in comparing results with Goldgaber, I wouldn't land in some sort of trouble.

The big winners of the amyloid gene finding, it was to be hoped, were all those who suffered from Alzheimer's. For now, at least, the gene connected to Alzheimer-type amyloid had been snagged, and convincingly so. Soon it was known that two other groups also had scored the gene—a collaboration between Colin Masters in Australia and Konrad Beyreuther and Benno Müller-Hill in Germany as well as the team of Henry Wisniewski, Nikolaos Robakis, and others at Staten Island's Institute for Basic Research. The four reports would be published in early 1987, with the German-Australian collaboration deserving special recognition. Lead author and graduate student Jie Kang had sequenced practically the whole gene—roughly 3,200 base pairs—while the rest of us had decoded a mere portion of it.

The new field of Alzheimer genetics essentially was christened. We had hold of a gene that looked suspiciously like a perpetrator—a candidate gene. Staring us in the face was evidence of a pathological pathway, from gene to protein to protein fragment to a miserable overabundance of amyloid plaques in the brain. But bear in mind that no one had yet spied a defect in the amyloid gene, a defect that might explain amyloid's insoluble pileup and Alzheimer's inheritance. We didn't even know if amyloid's pileup accounted for neuronal death. We had a gun, but no smoking gun.

In labs east and west the expectation floated that either one mutation, or perhaps several, in the newly captured gene might cause Alzheimer's. Like other groups, Neve and I carried out the chore of making sure the gene was active in the brain. Neve's delicate RNA tests revealed as much. Along the way we discovered—and we were the first team to do so—that the gene also was expressed in liver, heart, muscles, kidney, and many other type cells, which suggested that in its normal state it must play a major role throughout the body. But if this were true, why did its peptide build up and clump only in the brain? Before publishing our gene report, we also were trying to detect precisely where the gene lived on chromosome 21 by running it against markers on the 21 map.

Around Thanksgiving, some exhilarating news arrived from Peter Hyslop at Mass General. Scanning the breadth of chromosome 21, he and genetic analyst Jonathan Haines had picked up evidence of a gene flaw in the MGH families, particularly in the Italian family. The linkage wasn't terribly strong, but it was the first time ever that DNA analysis of an Alzheimer family had signaled the existence of an Alzheimer-associated gene defect. That, in itself, was enormous. The unknown disturbance lay in the vicinity of marker D21S1, the first marker Paul Watkins and I had isolated and plotted on our chromosome 21 map. Pulling out all stops and using D21S1's probes and the chromosome 21 map, I began looking to see if the amyloid gene lay anywhere near D21S1. If so, the defect Hyslop was picking up on might be located *in* the amyloid gene. Ergo, the amyloid gene truly might contain the answer to Alzheimer's. With more probing, one of mankind's supreme evils might be soon incarcerated.

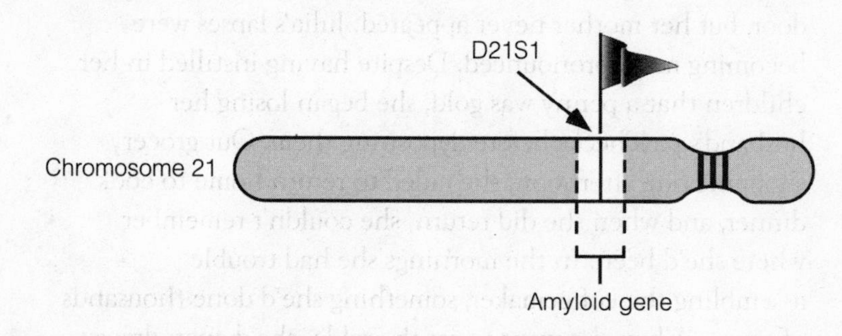

FIGURE 4.4 Amyloid gene's approximate location in relation to marker D21S1. Illustration: Robert D. Moir

Right before Christmas, we got a hint of positive data that—yes—the amyloid gene lay close to D21S1. Possibly on the verge of a fantastic finding, I redoubled my efforts. On both Christmas Eve and New Year's Eve I stayed in the lab until well past midnight, cranking out Southern blots to get a better reading of the gene's position on the chromosome. My absence at home didn't please my wife, and a hairline fracture crept into our marriage. It was true, the lab had most of my attention.

At last the mapping was finalized: the amyloid gene sat squarely within the very same area where Hyslop had signs of an anomaly. Very imaginably, somewhere along the gene's long corridor of bases sat the disease's fatal error. But I couldn't shake a disturbing image from my mind—the sight of that darn baseball skipping through Bill Buckner's hobbled legs into right field. What did it mean for our research? What did it portend for the suspicious amyloid gene and the purported linkage to chromosome 21?

One day Julia forgot to pick up one of her children at kindergarten. Five-year-old Julie, done up in coat and hat, had sat waiting and waiting for her mother by the school door, but her mother never appeared. Julia's lapses were becoming more pronounced. Despite having instilled in her children that a penny was gold, she began losing her husband's paychecks before depositing them. Out grocery shopping one afternoon, she failed to return home to cook dinner, and when she did return, she couldn't remember where she'd been. In the mornings she had trouble assembling the coffeemaker, something she'd done thousands of times. When she went to set the table, she'd open drawer after drawer before finding the silverware. Even more worrisome, she began to neglect the younger children—from diaper-changing to feeding. Once, without fail, she'd come into their rooms before bedtime to listen to their prayers. Now oftentimes she didn't. Her family didn't tease her anymore about her forgetfulness. What was happening to their mom? Where was her smile, her spunk, the optimism they'd always counted on? Why did she so easily grow sullen and slide into crying spells? Was she tired of them? Was it something they'd done? She gave no explanation.

— *five* —

Curious Gene

All the time the guard was looking at her, first through a telescope, then through a microscope, and then through an opera glass. At last he said, 'You are travelling the wrong way!'

—Lewis Carroll, *Through the Looking Glass*

It was easy to predict that the news scheduled to break in late February 1987 would cause a small explosion in the field, the media, and the public. The journal *Science* had accepted Rachael Neve's and my paper describing our isolation of the amyloid gene, and it was due to run alongside the Goldgaber-NIH team's similar accomplishment in the journal's February 20 issue. Our tests had gone only so far as to fish out the gene and map it to chromosome 21, not look for a mutation on it. But slated for the very same issue of *Science* was Peter Hyslop's report—with Jim Gusella as senior author—of an unidentified Alzheimer aberration that lay in the vicinity of the amyloid gene, which was bound to fuel the suspicion that the aberration occurred in that gene. It would be such a tidy explanation—that a mutated amyloid gene drove the disease by dumping too much of its protein into the brain.

Because of the map, probes, and blots I'd previously generated, I was second author on the Hyslop-Gusella paper. The chromosome 21 flaw it pointed to, one conceivably hidden in the amyloid gene, was scientifically sound, but something didn't quite add up. My 1984 probings of the Canadians' DNA hadn't shown a hint of a defect anywhere on chromosome 21. But maybe my tests had skipped over something. There was only one way to get a definitive reading of the gene—to use a variation in the gene as a marker and run genetic-linkage tests on the MGH's four Alzheimer families. Whatever it took, I wanted to secure an answer as soon as possible. It was too frustrating not knowing whether the gene was the instigator it appeared to be.

And so in early January I headed over to Gusella's lab with the purpose of diving into "Ye Old Blot Pile," as we called it. Stored on a shelf above one of the benches, it amounted to a big, bulging, white cardboard box piled high with nylon Southern-blot filters in which the four families' DNAs were immobilized. (The box was so bursting that eventually we renamed it "Ye Old Bloat Pile.") Reprobing the DNA bound to the filters should expose the amyloid gene's guilt or innocence. Gusella sometimes referred to me as a "free agent," because of the way I kept running between the lab at Children's Hospital, my designated grad-school workplace, and Gusella's Mass General lab, where I kept returning for reagents and advice, forcing a temporary relationship between the two. True to this alias, I gathered enough filters from Ye Old Blot Pile as well as from Peter Hyslop to permit a cursory scan of the amyloid gene.

My tests of the Canadians, the Russians, the Germans, and the Italians were finished by mid-January, and all I'd come by were starkly negative results. I'd held out hope—terrific hope, really—that the amyloid gene would yield a mutation in Mass General's four families, explaining their disease. But now that hope lay pretty much shattered. When I told Gusella and Hyslop that, while more comprehensive tests were needed, it looked like a mutant amyloid gene could be ruled out in the four kinships, I remember Gusella's reaction leveling me out a good deal. He was entirely unfazed. "Date your hypothesis, don't marry it," he as much as said. Hyslop sat silent. As usual, I couldn't read him.

Coinciding with the published accounts of the sightings on chromosome 21, a mid-February press conference took place at Mass General. The media piled into the Walcott Room, eager to know the lowdown. Did the findings that were about to break into print mean that a faulty amyloid gene was behind Alzheimer's disease? Were we on the precipice of an historic moment in the annals of medicine? Gusella, Hyslop, and I responded to the barrage of questions as best we could. To me, it was an incredibly awkward position to be caught in. As much as scientists attempt to deliver the truth, the traditions and ethics of science ironically can force white lies. I couldn't yet communicate what I had preliminary evidence of—that the gene that everyone was thinking *might* hold an Alzheimer mutation actually appeared mutation-free in the four families—because the data needed stronger confirmation. Consequently, I did my best to dodge inquiries about the gene's culpability. Paul Watkins was in the audience, and I remember him wearing the slightest of grins. He knew the latest word on the fished-out gene; he knew how much I couldn't say and how much I was squirming.

As expected, the February 20 issue of *Science* created a global swirl of interest. Accounts by the German-Australian team and the Staten Island group of their separate amyloid gene isolations appeared around the same time in *Nature* and the *Lancet*, respectively, adding to the surge of media interest. In his comments to the press, Gusella tried to downplay the idea that the reported anomaly lay in the amyloid gene, ergo the gene triggered Alzheimer's. Five hundred or so other genes occurred within the same region of chromosome 21, he pointed out. Any one of them might be home to the unidentified flaw detected in the Mass General families.

Major newspapers nonetheless rocketed out the story with headlines that tended to be overly conclusive: "Scientists Identify Gene for Alzheimer's"; "Cause of Alzheimer's Disease Found." In the wake of the news, more than one press photographer asked me to peer into a microscope as if eyeballing the retrieved amyloid gene; as if we had proof of its guilt; as if genes could be seen under a microscope. Despite not sharing in the gene's isolation, from his lab in La Jolla George Glenner cheered on the field's progress. "These reports," he told the *San Diego Union-Tribune*, "completely confirm what we had predicted [that the amyloid gene lay on chromosome 21] and I am delighted at the speed with which these developments have come."

In the near future, at a gene-naming contest held at a human genome meeting in Paris, I would offer the suggestion that in keeping with the name Glenner and Wong had given their retrieved brain fragment—the beta protein—why not name the parent protein it sprang from the amyloid beta-protein precursor, since it was the precursor of the little snippet that landed in plaques. For both gene and protein, this name was

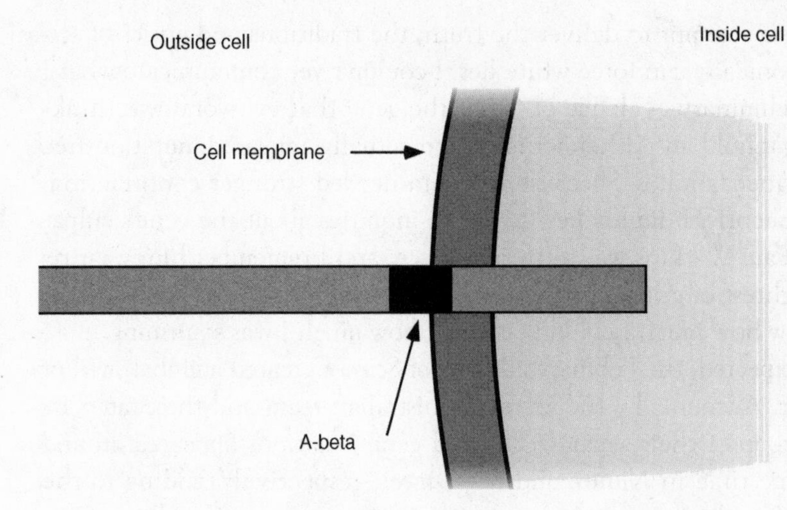

FIGURE 5.1 APP protein and its A-beta segment sticking out of a cell. Illustration: Robert D. Moir

adopted, at least temporarily. Today, gene and protein are both referred to as "APP," for *beta-amyloid precursor protein*. As for the APP protein's plaque-embedded fragment, it would soon go by the official nickname of "A-beta," for *amyloid-beta*.

For a brief period, the field was churning over the possibility that a flawed APP-amyloid gene provoked inherited Alzheimer's. On my own part, I felt hopelessly confused. Logic suggested this suspicion was true. Look at how patients' brain tissue was glutted by amyloid derived from the APP protein. And here Hyslop had linked the disease to the very area the APP gene sat in. Yet a fuller analysis of the MGH's four families had confirmed that APP held no mutation, at least not in those families. A collaborative study of two Belgian early-onset Alzheimer families by John Hardy in London and Christine Van Broeckhoven in Belgium reached the same conclusion, both their report and ours airing back-to-back in *Nature* later in 1987. As our title put it, "The genetic defect in familial Alzheimer's disease is not tightly linked to the amyloid B-protein gene." I had to wrangle for that title as *Nature*'s editors wanted something more provocative and conclusive along the lines of: *APP is not the Alzheimer's gene*. "But we don't know that," I told them. "Around the world there are scores of other Alzheimer families in which a mutation on the amyloid gene hasn't been ruled out yet."

If the four families had no mutation on APP, how was it they processed too much brain amyloid? What was going on? Quite possibly another gene, even one on a different chromosome, held a defect and worked at cross-purposes with APP, forcing its production of amyloid. A tree falls against another tree, which falls and does major property damage. (There was a growing sense that more complex genetic diseases might be the result of several off-kilter, interacting genes.) No one was yet ready to rule out APP's involvement in Alzheimer's. It was, after all, the source of Alzheimer's runaway brain amyloid.

In March of '87, an altogether different mode by which the APP gene might be causing Alzheimer's had the field buzzing anew. A few years earlier, Miriam Schweber, a biochemist at Boston University, had struck out to try to fathom a molecular connection between Alzheimer's and Down syndrome. She was curious as to why Down's individuals developed not only Alzheimer's brain markings, but also in some cases its brand of dementia—an observation scientists had made all the way back in the late 1940s. Dementia, which develops over time, can be distinguished from mental retardation, which one is born with.

To what degree those with Down's suffer the effects of Alzheimer's dementia remains highly controversial, observes Charles Epstein, a molecular geneticist at the University of California, San Francisco, and an authority on Down syndrome. Differences in diagnostic criteria make it hard to arrive at accurate numbers. "Although Down syndrome people have Alzheimer's pathology by the fourth decade, the prevalence of dementia is certainly less than that. For Down's subjects over age fifty, most studies suggest the prevalence is on the order of 40 to 50 percent," notes Epstein.

Schweber reasoned that if Down's people got Alzheimer's, Alzheimer's similarly might spring from an extra copy of chromosome 21 DNA. Devising a technique for measuring DNA, she flushed out what she believed was tantalizing evidence: a tiny stretch at one end of chromosome 21 appeared to be triplicated in Alzheimer patients. But when other scientists tried to duplicate her fining, none reported being able to do so.

Now that the isolated APP-amyloid gene had all the allure of an Alzheimer's suspect, two collaborating groups decided to borrow from Schweber's theory and check to see if APP just might be erroneously triplicated in patients. One of the scientists was my colleague Dmitry Goldgaber at NIH. Since our earlier skirmish, we'd been in touch now and again, and seemed on okay ground. Goldgaber had joined forces with Jean-Maurice Delabar's lab at Necker Hospital in Paris, and, as I recall, it was from Paris that Goldgaber called me one day to share the news. They had confirming evidence from brain DNA! They'd found a triplication of a portion of chromosome 21 that contained the APP gene. If their observation held up, it would be a stunner. Perhaps as a fetus developed, an abnormal tripling of genes paved the way for Alzheimer's.

But their data, which had been culled from only three cases of sporadic Alzheimer's, quickly toppled. Rachael Neve and I were unable to turn up the same finding. These negative results, which I communicated to Goldgaber, didn't make him too happy, since his and Delabar's report of triplicated DNA in patients was on the verge of appearing in *Science*. Published in March 1987, their paper caused a media groundswell: *Alzheimer's Might Be Caused by Extra Gene*. But the following October, Neve's and my paper, a second one by Hyslop and me, and a third by Marcia Podlisny and Dennis Selkoe appeared in a row in *Science*, each refuting the triplication theory. The media's echo followed: *Alzheimer's Not Caused by Extra Gene*.

In a very short time, then, two possible scenarios for how the APP-amyloid gene might drive early-onset Alzheimer's had bitten the dust. In the families tested, it didn't look as though the gene contained a relevant mutation. Nor that an extra copy of APP explained the disease. At Children's, both Rachael Neve and I nevertheless were riveted to the newly captured gene and submerged in tests that might expose more of its characteristics.

The aim of my Ph.D. thesis—to characterize APP and show that it was connected to Alzheimer's causation—necessitated this concentration. I could forget proving that a mutated APP single-handedly caused the disease in the Mass General families. But excluding that hypothesis seemed equally valuable. And I wanted to keep gnawing away at the puzzle that APP presented, for it truly seemed that APP must be a critical accomplice in the disease, given that the A-beta peptide—a mere snippet of APP's full protein—was at the core of Alzheimer's copious plaques. Perhaps other genes and their proteins forced APP to release as much A-beta as it did.

So many things about the A-beta fragment stumped investigators. How did it end up in plaques? Did it get released from live neurons or dead ones? Many researchers suspected that *proteases*, enzymes that cleave proteins, set A-beta free from its parent protein. The biggest question of all was whether A-beta and the plaques it aggregated into brought death and ruin to neurons.

First things first, Rachael Neve and I tried to pin down which cells made the APP protein, and once processed inside the cell, where it traveled in the cell. In the early spring of '87—around the time the triplication theory was attracting attention—we were poking along various segments of the APP gene when we realized that one of the plucked-out APP genes we were inspecting had a short extra sequence of DNA stuck in its middle. Quite possibly this insert was a contaminant that our genetic engineering procedures had accidentally introduced. Yet tests showed it was a legitimate part of APP, although only occasionally present in APP made in the brain. It was a wholly unexpected, thrilling find—much like suddenly discovering a room in your house you never knew existed! Perhaps this oddball DNA sequence, which was all of 168 bases long and encoded only 56 amino acids, would shed light on the clip-and-clump puzzle—how the A-beta shard got clipped free from its bigger protein, allowing it to clump into amyloid deposits. The APP

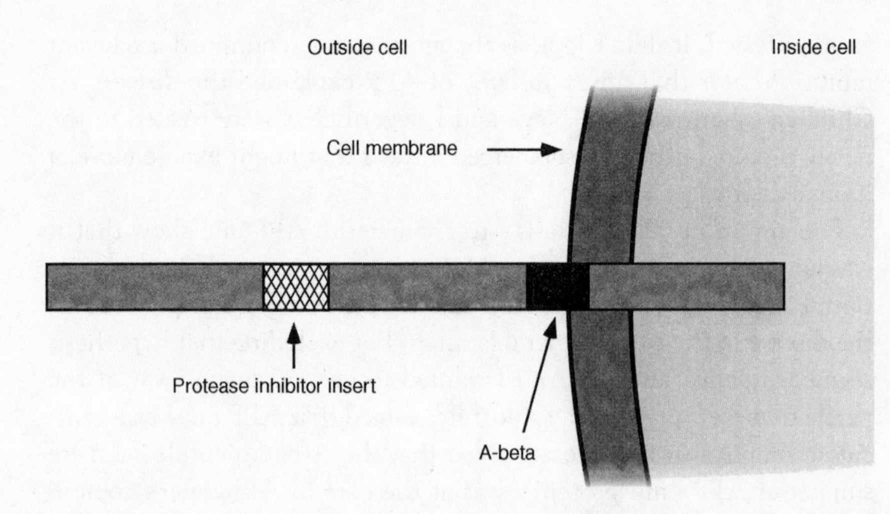

FIGURE 5.2 APP protein and its protease inhibitor domain.
Illustration: Robert D. Moir

gene and protein having been so newly found, everybody was looking for immediate answers. Maybe we'd stumbled on one.

Andi McClatchey, a technician in Jim Gusella's lab who was proficient in the art of manually sequencing DNA, had been helping Rachael Neve and me with APP sequencing work. She agreed to decode the newly found insert as well. It was around this time that I was preparing to leave the lab at Children's, which by then was directed by Neve, and head back to Mass General. An initiating factor was that David Kurnit, my rotation adviser, was relocating to the University of Michigan at Ann Arbor. Kurnit approved my return to Gusella's lab, although initially I didn't seek clearance from Harvard's higher-ups. Admittedly, I broke a few Crimson rules. During my sixteen months under Kurnit, I'd been immersed in either isolating or studying the APP gene and had managed to avoid the full complement of lab rotations usually required of grad students. And now I was leapfrogging back to a lab I'd already worked in. Trusting I'd get an official okay, I asked Gusella to take over as my adviser and sounding-board for my thesis.

When I reentered Gusella's fold in the spring of '87, it felt good to be back home in the thick of things. Gusella, together with John Growdon, the director of Mass General's Memory and Disorders Unit, had acquired many more Alzheimer families for analysis, not to mention DNA samples related to a widening swath of other inherited neurological disorders. With over 10,000 cell lines stored in its liquid-nitrogen freezers, Gusella's "Neurogenetics Laboratory" was on the verge of broadening into the "Molecular Neurogenetics Unit," and well on its way to becoming one of the country's most active centers of human genetic research. Surprisingly, even with an international army of scientists giving chase, the Huntington gene on chromosome 4 hadn't been cornered and conquered yet, causing no uncertain frustration.

That summer, just as Andi McClatchey started decoding APP's odd little insert, I departed for a human genome conference in Paris. Ever since Rachael Neve and I had isolated the APP gene, invitations to attend meetings and discuss any known particulars about APP had been flowing in from different continents. It's strange. I'd always dreamed, in my band-playing days, of traveling to distant shores, and now this other profession was allowing that dream to come true. Pursuing science was surprisingly similar to playing music, I was discovering. Behind closed doors, drawing on creative, introspective juices, you generate the data. On tour, hitting one new stage after another, instead of playing set-lists of songs in front of noisy audiences, I found myself presenting sets of slides to much quieter but no less rapt listeners. The stage being familiar ground, scientific presentations to large audiences were that much easier; I wasn't exactly mike-shy. It was a relief not to have to worry that people weren't getting up to dance—this always having been a big concern for my band-mates and me. But I did have to worry if my talk, upon completion, produced dead silence with no questions asked. It could be a sign that my data were incomprehensible, or worse, comprehensible but ludicrous. On the other hand, an informative talk that carried crystal-clear implications also could raise little or no discussion.

Jim Gusella had crossed the Atlantic for the same Paris meeting, and one afternoon, both of us having ducked out of the conference to amble along the Seine, we ran into each other.

"So did Andi finish sequencing the APP insert yet?" I asked. We'd decided to cross the next bridge and continue down the Left Bank to visit the Musée D'Orsay and its Impressionist masterpieces.

"She did. She got something," Gusella said, somewhat absently. Knowing how his mind could peel off in any direction, I half expected him to change topics and launch into some obscure trivia dealing with the French Revolution.

"So what did she get?"

"Oh, just some novel sequence."

I began nudging his shoulder, not entirely averse to the idea of pushing him into the Seine since he was being so recalcitrant. "Did she check the database? Did the sequence match anything?"

As soon as you sequence what might constitute a newly found gene or protein, or piece thereof, the usual routine is to type out the sequence into a computer file and plug it into GenBank, the giant database maintained by the National Center for Biotechnology Information, a division of the National Institutes of Health. This can tell you if any similarities exist between your sequence and GenBank's huge roster of previously recorded sequences. Established in the early 1980s and daily exchanging information with two similar databases, one in Japan and one in Europe, GenBank's computerized vault essentially holds nearly every stitch of DNA and every stretch of amino acids that scientists have so far extracted from Earth's three major categories of life-forms: *archaea* (ancient bacterialike organisms that reside in extremely hot places, like hydrothermal vents); *bacteria* (or prokarya); and *eukarya* (organisms with cells that have a nucleus, from yeast to humans). Therefore, if GenBank turns up a DNA or protein match to your sequence, or even a partial match, sometimes it can tell you volumes about your unearthed sequence's evolutionary history, its function, or even its location in a genome.

When the APP gene was found in the fall of 1986, it hadn't matched an iota of anything in the database. In the years to come, however, as more sequences were added to GenBank, we'd see that likenesses of APP occur in species up and down the evolutionary ladder, save for yeast and bacteria. Even fruit flies and nematodes have an APP-like gene, suggesting that as earth's species evolved away from the ancestor of all ancestors—a speck of a microbe—an APP-like molecule has been conserved in most species because it probably attends to an important life function. Currently, GenBank lists nearly 5 million DNA sequences obtained from some 41,000 species—the corresponding protein sequence being an implicit part of each listing. But back in 1986, fewer than 16,000 sequences had been submitted. These included, as holds true today, far more bits and pieces of genes than complete genes.

In the case of APP's little insert, I was guessing that since it was only occasionally present in the APP protein made in the brain, it must have a very specific function, which a similar GenBank sequence might shed light on.

"So what did Andi find? Did the insert match anything?"

"It matched a cow protein or something like that," Gusella replied.

"Cow?"

"Yeah, it matched a bunch of protease inhibitors, some from cows, some from snakes."

"Protease inhibitors!" I couldn't believe how impassively Gusella was telling me all this. After all, one of the most tantalizing puzzles about the APP protein was, What clipped A-beta loose from its large protein? What set it free for its rabble-rousing into plaques? The answer was thought to be a protease, a type of enzyme that snips other proteins much like restriction enzymes snip DNA. (A protease is simply a specialized enzyme, an enzyme a specialized protein.) A protease inhibitor, conversely, inhibits the cutting action of proteases. If APP sometimes contained a protease inhibitor, it could be good or bad. On the good side, it might be blocking proteases that free amyloid-forming A-beta. On the bad side, it might be inhibiting beneficial proteases that destroy A-beta before it glomps into plaques.

Gusella, as it happened, was entirely attuned to the significance of this finding. He was just getting his chuckles by downplaying it and making me drag it out of him.

When I got back to Boston and took a look in GenBank, likenesses to the human APP insert included protease inhibitors from a turtle, a cow, a viper's venom, and even an "edible" snail, as listed. If a similar sequence had been conserved throughout the ages in an organism as ancient as a snail, more than likely the little insert in human APP indeed served an important purpose. But what, exactly?

```
Human ···EVCSEQAETGPCRAMISRWYFDVTEGK APFFYGGCGGNRNNFDTEEYCMAVCG···
 Snail ···SFCNLPAETGPCKASFROYYYNSKSGGCQQFIYGGCRGNQNRFDTTQQCQGVCV···
```

FIGURE 5.3 The shading indicates amino–acid commonalities between the APP protease inhibitor in humans and snails. Illustration: Robert D. Moir

I look back on the finding of APP's protease inhibitor as one of the most exhilarating I've ever lent a hand in. As is par for the course in science, it ended up that two other teams had stumbled onto APP's insert as well. I found out about them several months later, in February 1988, when, on the eve of the paper authored by Neve, Gusella, and me describing the insert, an Australian reporter dropped mention we'd be sharing space in *Nature* with similar accounts. A bigger surprise would descend at the 1989 NATO Advanced Research Workshop in Maratea, an Italian beach resort south of Naples. A protein biochemist I met there—Steve Wagner, who is now a close friend—told me that the APP insert was the same little protease inhibitor that his lab at UC, Irvine, headed by Dennis Cunningham, had isolated in 1987 and determined to be associated with blood coagulation! I hadn't cross-referenced the protein side of GenBank, which GenBank automatically does now, so this had escaped me. The Irvine group, meanwhile, had been unaware that the protease inhibitor was from the APP gene. Thus emerged through the synergy of science a better understood little molecule.

But not altogether understood. Even though the discovery of APP's extra little piece caused great excitement throughout the field, studies initially failed to reveal a role for it in the disease. (Just lately we have evidence that elevated inhibitor-type APP *is* tied to surplus amyloid and the disease, providing drug researchers with a tempting target.)

As the most frequently mouthed phrase in science goes—more research was needed. In myriad labs, as a swarm of researchers tried to shake out insights into APP and its place in Alzheimer's, all of us were being reminded for the umpteenth time that little in science is quickly understood.

The Noonan family kept thinking that whatever was troubling Julia would pass, but it didn't. It only grew worse. Where once she never was off her feet, now she sat idle for long periods with a hollow expression. The family doctor believed her symptoms pointed to depression and prescribed antidepressants. But pills had no effect. John Noonan was in such turmoil over his wife's unexplained behavior, he had a hard time keeping his mind on his job as firefighter; managing the household and ensuring that the children at home were looked after was beyond him. The very youngest went to live with married siblings, and the older ones at home began filling their mother's shoes and raising the others. Doctors recommended shock therapy, and in the spring of 1967 Julia underwent two series of treatments. These too had no effect. That December she was admitted to the psychiatric unit at St. Elizabeth's Hospital outside Boston for examination, and a neurologist shared her opinion with the family. Julia either was schizophrenic or had a very rare disorder called Alzheimer's. It was Christmastime. Julia came home for a spell before entering a nursing home. A wreath hung on the Noonans' front door, and under the tree lay presents that John Noonan had bought and wrapped on his own. But for him and the flock of children that gathered, Christmas was as empty as the look on Julia's face.

— *six* —

From Famine to Feast

Magnify the small, increase the few.
Reward bitterness with care.
—Lao Tsu, *Tao Te Ching*

By the end of 1987 the once compelling notion that the APP-amyloid gene contained a mutation that triggered heritable Alzheimer's lay pretty much dead in the water, at least for the time being. It was the beginning of a fairly frustrating and confusing interlude for the field, as well as for my own work. APP seemed involved in the disease's mechanism, but we had no idea how. Other genes might be involved, but we didn't know which ones. Generally, a contagious interest in the neurosciences was attracting researchers by the scores into various sectors of Alzheimer's research. As a result, where once we had scant leads to go by, now, because of multiplying hands, there was a dizzying number of disparate clues. From chaos, one had to hope, would come order.

Mass General's original statistical red flag that had indicated something amiss on chromosome 21 still stumped investigators throughout the field. Since APP didn't appear to be the transgressor, growing numbers of scientists were embracing the theory that quite possibly the hidden defect resided in a different gene on chromosome 21. And perhaps this other gene's mismade protein interacted with the APP protein, resulting in APP's released A-beta fragment. Thereupon the crowded ferry went from listing heavily to port to nearly as heavily to starboard as numerous geneticists shifted from one railing to another to spy the newly rumored leviathan.

My thesis meanwhile kept me trying to shake out insights into how the APP gene played into "the amyloid story," as George Glenner called it. This amounted to setting up comparisons of the APP protein in healthy people versus Alzheimer patients. In roughly a year's time I'd contributed to seven papers in *Science* and *Nature*, four of which I'd first-authored. Most of these reports pertained to APP, and when several were highlighted in the publication *Current Comments* as among the most cited papers in the life sciences in 1987 and 1988, it was largely a reflection of the enormous surge of interest in APP. Former skeptics were fast becoming die-hard believers in the concept that genes that led to freewheeling A-beta and plaque formation must be involved in the disease process.

Due to my APP work, lecture invitations and travel opportunities kept beckoning. Everywhere I went I heard about the efforts of researchers based from Florida to Seattle, London to Florence, who were hunting for the hypothetical "other" gene on chromosome 21, a mutated specter that so far had yielded no clear traces. Twenty-one's midsection "was under the microscope," remembers Paul Watkins. "Lab groups around the world were focusing all their attention on trying to clone genes out of it." With any luck—*the* gene.

My overseas talks in 1988 brought experiences I never in my wildest dreams imagined befell a scientist. In conjunction with an international Alzheimer symposium in Helsinki, those of us in attendance were dazzled by a private performance of Sibelius by the city's symphony orchestra. At a reception in Tokyo tied to an Alzheimer conference, each male was presented with his own gracious female escort, who went to disconcerting lengths to fill our glasses after each sip, causing my then-wife, Janet, to wryly wonder why she hadn't been given her own male equivalent. A trip to New Delhi, where I gave the keynote address at a World Federation of Neurology conference, brought the unsettling experience of riding atop an elephant to the opening banquet—Janet in evening gown and myself in black tie—while passing the most abject poverty we'd ever witnessed.

Through travel came increased exposure to many of the Alzheimer field's top brass and "road-show" performers, researchers I would increasingly make friends with and collaborate with, and at various other times cross swords with and race against. One particularly memorable bonding episode happened, of all places, in a sauna on Hanasaari, a small island off Helsinki where a symposium on the pathobiology of Alzheimer's was in session. We were given heated wet birch branches and told by our hosts that if we whacked ourselves repeatedly on the back, it would do wonders for the circulation. It felt like a weird, strange dream to be sitting there buck naked with several of the field's senior statesmen—who will remain nameless to protect their modesty—and engaging in what seemed like a sadomasochistic tradition. The scene took an even zanier turn when the branches started flying and a circle of grown men, pale as cooked veal and some quite paunchy, starting hitting each other with total abandon and sheer glee—a circle of scientists gone up in steam.

It was during a second trip to Japan in 1988 that I first met George Glenner. Lugging a big old brown leather suitcase, he was introduced by another colleague as the three of us boarded a bullet train at Narita airport for downtown Tokyo. The encounter was brief, little more than a greeting and a handshake, but it was a truly great moment to finally meet the man without whose peptide achievement, and its timeliness, I mightn't have been in Tokyo or hell-bent on wresting out components of Alzheimer's "amyloid story." His smartly dressed appearance, refined features, and mellifluous voice left the impression of a stylish older Cary Grant, not a fellow who studied brain tissue. At the time, he was still trying to work out a method for measuring the A-beta peptide that could be used to diagnose Alzheimer's. But he wasn't having much luck finding a consistent pattern of A-beta in the blood of patients.

One consequence of the concerted search for the "other" gene on chromosome 21 was that our Mass General 21 map was all the more in demand. Somewhat belatedly, in the spring of '88 it even was headed toward a formal debut in the journal *Genomics*. All told, chromosome 21 was turning into one of the best documented strands in the genome. Labs had banded together to plot mile-markers along it, and it had become a crown jewel representative of a low-budget, hands-on era of genetic mapping, soon to be overtaken by elaborate technology that would immeasurably speed the mapping and sequencing of many genomes—yeast, worm, fly, mouse, horse, human, and countless others. The same machine operating off a different program would similarly revolutionize genotyping, the process of determining which two versions, or alleles, of a gene a person inherits. As invaluable and time-saving as the new technology has proven, today I sometimes yearn for the old days in Gusella's lab when, in order to genotype DNA, we would fill baggies with blotted DNA, add radioactive DNA probes, squeeze out the bubbles that potentially could mar our results, wash the filters in tubs, lay them out by hand, dry them and press them onto X-ray film, then wish them well before setting them overnight in a freezer at –80 degrees Celsius.

On the day of reckoning, in the red sheen of the darkroom, we'd eagerly await the developing image. After weeks of work, as the image of DNA bands emerged on film there was an excitement I can hardly describe. Today, the glory of accomplishment gained from this gradual process has lost ground to getting results from a computer lickety-split. Automation has been tremendous for science, but I worry that knowledge has been lost along the way. I see too many postdocs who, spoon-fed by computerized machinery, have little sense of the intermediary steps involved.

From the meetings I was attending, it was easy to see that research into every imaginable crevice of Alzheimer's pathology was expanding at a furious clip. The total land-sea-and-air assault included mounting numbers of molecular biologists, chemists, neurologists, pathologists, microscopists, psychologists, neuroimmunologists, as well as geneticists, who were increasingly banding together, their interdisciplinary approaches promising all the more headway into the disease's depths. Studies were imparting new details about the immune response that attracts swarms of housekeeping glial cells to amyloid plaques and dying neurons to clean things up, and about how this glial attack possibly worsens cell death; about mitochondria, the energy providers inside cells that possibly contain inherited defects associated with Alzheimer's; about the selective vulnerability of neurons in specific brain regions; about destabilized synapses between brain cells; about plaque-embedded proteins other than A-beta whose genes possibly contributed to the fracas.

Progress even was being made into the impervious tangles inside cells—the disease's hallmark lesion that had been so difficult to penetrate. Primarily by staining tangles with antibodies, researchers who had doggedly stuck with this lesion were realizing that the tangle subunit resembled a modified form of a protein called *tau*. In its normal soluble state, tau is essential to a cell's infrastructure. In essence, it forms the cross-beams for a cell-wide span of railway tracks—filaments called microtubules—which are constantly laid down and taken up for the purpose of shuttling types of protein cargo around a cell. The tau in tangles, however, appeared to be a twisted, highly insoluble version of tau. Corrupted tau might mean the kiss of death for a cell, and tangle researchers were becoming all the more dedicated to the idea that perhaps the tangles after all represented the lesion most closely associated with Alzheimer's massacre of neurons. Tau's gene had been identified in 1987, but similar to the APP-amyloid gene, no Alzheimer mutations had been spotted in it.

Research by Peter Davies at Albert Einstein College of Medicine was providing the lasting value of revealing that tangles were only "the tip of the iceberg of a very widespread tau-related abnormality" that begins far earlier in the disease process, as Davies today notes. This phenomenon was noticeable in tangle-free neurons, thus perhaps the beginning signs of tangle formation. As tangle research intensified, occasionally I'd wonder if we were wasting our efforts on amyloid. But these doubts were fleeting. A gut feeling told me that amyloid was more germane to the disease.

Developments in the field were accelerating so rapidly that a special meeting on the molecular mechanisms of Alzheimer's disease convened in April 1988 at Cold Spring Harbor Laboratory. It was the third Banbury Conference in five years to address work in our field. Gatherings at Cold Spring Harbor, one of the country's oldest bastions of molecular science, had garnered a reputation for being where the action was. It was here on Long Island's north shore that James Watson, in 1953, gave the first public report of his and Francis Crick's completed model of DNA's structure. Touching the sky at Cold Spring Harbor remained the standard, as did discussions over controversial topics. Attending the meeting were Henry Wisniewski, Dennis Selkoe, Colin Masters, Konrad Beyreuther, Peter Davies, Zaven Khachaturian, and other stars of the Alzheimer's Road Show, although three of the field's most prominent—George Glenner, Robert Katzman, and Robert Terry—were absent.

Nearly half of the talks to be presented involved studies of the APP-amyloid gene, showing how fixated the field had become on the puzzle of amyloid's role in Alzheimer's. Nonetheless, one of the conference's highlights was tangle work done by Michel Goedert and Aaron Klug from the Medical Research Council Laboratory of Molecular Biology in Cambridge, England. Four years had passed since George Glenner's isolation of what proved to be the plaques' core peptide, and now this English team announced it had biochemical proof that the core protein in Alzheimer tangles indeed *was* contorted tau, the protein which in a healthy, soluble state was so crucial to a cell's infrastructure. Tangle researchers were elated. Until then, tau had been fingered, but not pointedly singled out. Since tau was key to a cell's survival, here was grist for the argument that the tangles more than the plaques might figure in the disease's neuronal havoc.

On the second day of the conference, one of the field's more famous confrontations occurred. It had little to do with the developments we'd gathered to discuss and was more a poignant reminder of how personal histories can influence scientists' perceptions of and approaches to science. While most Banbury talks and their follow-up discussions in those days were printed in book form for posterity, the words delivered and exchanged during this particular session would be considered inappropriate for publication and omitted.

Benno Müller-Hill took the stage. A molecular geneticist from the University of Cologne, Müller-Hill was acclaimed for his role in initiating Germany's postwar molecular research efforts as well as his work on how genes got turned on and off in bacterial viruses. More recently, his collaboration with Konrad Beyreuther and Colin Masters constituted one of the four teams to have isolated the APP gene. Unlike the rest of us, they'd gone the full mile and deciphered the gene's complete sequence. I was especially curious to hear Müller-Hill speak, because the night before he'd told me he was going to say some things that no one would soon forget. I assumed he meant new inklings about APP.

But Müller-Hill—in his midfifties, scraggly-haired, and in garb that furthered the image of an aging hippie—made surprisingly short shrift of his group's major APP accomplishment. Instead he moved on to express his great concern over the rapid-autopsy programs under way in the United States. Usually an autopsy is performed and brain tissue fixed within approximately ten to twenty hours of a person's death. Rapid-autopsy protocols, on the other hand, aim for this procedure to happen within two to four hours of death so that tissue will be preserved in its freshest state for the clearest possible reading of its pathology before postmortem enzymes markedly degrade the tissue. For some—Müller-Hill plainly among them—it conjured up the inhumane image of an attendant tapping his foot, waiting for someone to die so that the brain could be promptly snatched away. In the audience, listening with growing fury to Müller-Hill, was Allen Roses, who was associated with a rapid-autopsy program at Duke University Medical Center, where he served as chief of neurology.

Rapid autopsy reflected a worrisome disrespect for life, Müller-Hill somberly stressed, a numbing-down of moral virtues and a gradual loss of reverence for human life that was reminiscent, in his opinion, of what had transpired in the early evolution of Nazi Germany. We should worry that rapid autopsy's rush to the brains of the dead was taking us down a similar path. It wouldn't be long before physicians would have gas pumps full of formaldehyde at the bedsides of the dying, he grimly joked, so as to be able to preserve brain tissue the second after a person's last gasp. Müller-Hill's book *Murderous Science*, an examination of how government and science converged to give rise to Nazi atrocities, had been published a few years earlier, so his statements weren't just coming out of left field.

The longer Müller-Hill expressed his views, the more upset many listeners became. The inferred parallel between Nazi science and rapid-autopsy procedures "was very unfair," remembered Henry Wisniewski. "Rapid-autopsy was a noble activity to help people" by way of research. Allen Roses left the room, and when he returned he was visibly shaking. Addressing Müller-Hill, he unleashed a sharp rebuke. The words I recall him saying were along the lines of: How dare you associate me with the butchery that took the lives of my people! Others in the audience were equally outraged. By the time Müller-Hill's talk ended, some were so fired up they seemed on the verge of leaping on stage to attack him.

Müller-Hill later told me that everything he had said and done had helped him do his job. His old clothes, his disheveled state, his inflammatory words—they all helped create a drama his listeners would always associate with the point he wanted to make. No one would forget his warning. I couldn't help but be struck by this highly questionable yet selfless act. He felt so passionately about getting his message across that he'd been willing to risk his reputation in a field that strongly ascribes to external validation—what our colleagues think of us. Several reports have since been published contending that postmortem brain tissue doesn't degrade as quickly as once thought. Therefore the final word on rapid autopsy may be that it's simply not necessary.

In the spring of 1988 I received some bleak news. My thesis advisory committee, which included various distinguished professors from inside and outside Harvard, decided that my goal—to demonstrate that the APP-amyloid gene was instrumental in Alzheimer's pathology—was lacking because APP didn't appear to harbor a mutation that made it *the* disease gene. As Robert Horvitz and other members of the committee advised, "You've shown us what the Alzheimer's gene is not," as opposed to what it is. Horvitz, celebrated for his probings of the microscopic worm *Caenorhabditis elegans*, was a master molecular biologist, an MIT professor you were blessed to have on your thesis committee, but he had a reputation for being extremely demanding of grad students. My response to the committee was, "I thought the aim of classic scientific method was to exclude hypotheses. I cloned APP, characterized it, and ruled out the theory that it causes Alzheimer's." But this rebuttal convinced no one. It was decided that I'd have to pick another Alzheimer candidate gene, isolate it, and run the same linkage tests all over again. Worst of all, I wouldn't be able to graduate in the spring of 1989 as I'd hoped, but if I surmounted the new task set before me, I'd receive parchment a year later. It seemed as though my bumpy road on the way to a doctorate was going to remain bumpy until I graduated.

For my new project I chose a family of three heat-shock genes, one of which purportedly was on chromosome 21 and a prime suspect in terms of harboring an Alzheimer mutation. Thus, in a roundabout way, I found myself becoming very involved in the hunt for the "other" gene on 21, which no one had any new clues about. Heat-shock genes are so named because the proteins of those in bacteria get activated in very hot conditions. In the human body, they protect cells from harmful protein by-products that can arise from a high fever, toxins, or other stresses. I hypothesized that a mutated heat-shock gene-protein facilitated the apparent miscutting of the APP protein, leading to excess amyloid and, in turn, Alzheimer's disease. Working away in Gusella's lab, by the fall of '89 I'd isolated and mapped all three heat-shock genes. Not only did they not link to Alzheimer's, none even resided on chromosome 21, in contrast to what had been previously reported. So there I stood, as empty-handed as before. I still hadn't found a gene mutation connected to the disease's origins, the original intent of my thesis, but I had to hope that there was no way in good conscience my thesis committee could hold me back again.

I was so frustrated about having to jump through this additional hoop for my Ph.D. that I never produced a formal journal account of the heat-shock findings. Slowly writing up my thesis, I straightaway resumed my investigations into APP—particularly into how the A-beta piece of APP's protein ended up in the disease's plaques.

While attending a neuroscience meeting in La Jolla earlier in 1989, I'd paid a surprise visit to George Glenner, and as we stood in a courtyard outside his UCSD lab—Glenner, I couldn't help but notice, liked his smokes, going through several Marlboro filters in quick succession—we'd discussed what a terrific feat it would be to deduce exactly how the A-beta snippet was set free from its parent protein. Then one might target that mechanism with a drug and stop amyloid's buildup in the brain. I'd shared my bewilderment over the absence of mutations on APP, which by every right, it seemed, should exist but hadn't surfaced. Gentle and avuncular, Glenner had advised, "Stay totally focused on APP; it's a means to an end."

It was hard to keep track of all the new pieces of evidence flowing out of labs and harder still to figure out how each piece fit the puzzle of the disease's pathology. In particular, "the explosion of research" into Alzheimer's amyloid was providing researchers with "almost an embarrassment of riches"—this fitting commentary by Dennis Selkoe appearing in *Science*. Highlights of the steady progress included new hints that the APP protein normally helped nourish and sustain connections between neurons. So abnormal APP very well might be a death knell for the brain. Of great intrigue was a finding made by Bruce Yankner and my former teammate Rachael Neve at Children's Hospital that suggested that when released from cells, parts of the APP protein might be capable of becoming toxic. If this could be more convincingly shown in experiments with neurons, it could amount to the first hard evidence that abnormalities associated with APP could desecrate the brain's neural circuitry.

Very tellingly, too, various labs were piecing together how the sticky A-beta fragment came to be lodged in Alzheimer's brain plaques. Once made deep inside the cell, APP's full protein got transported to the cell's outer membrane. This made it a "membrane" protein, as opposed to proteins that float around inside the cell or are exported in their entirety outside the cell. And in the membrane it got stuck, much like a robber trying to squeeze through a window, as Dora Kovacs, a cell biologist in my unit, describes it. "Part of the robber dangles into the fluid outside the cell, part of it dangles into the fluid inside the cell, all due to the fact that these exterior and interior parts are more chemically at rest in water," notes Kovacs. "The middle part is repelled by water and therefore stays in the membrane."

As illustrated below, the little A-beta portion of APP, which is located in the robber's upper midriff, extends from inside the membrane to just outside the cell.

It was assumed that like many membrane proteins, some part of the robber's outer portion got clipped free by an enzyme, liberating it to go

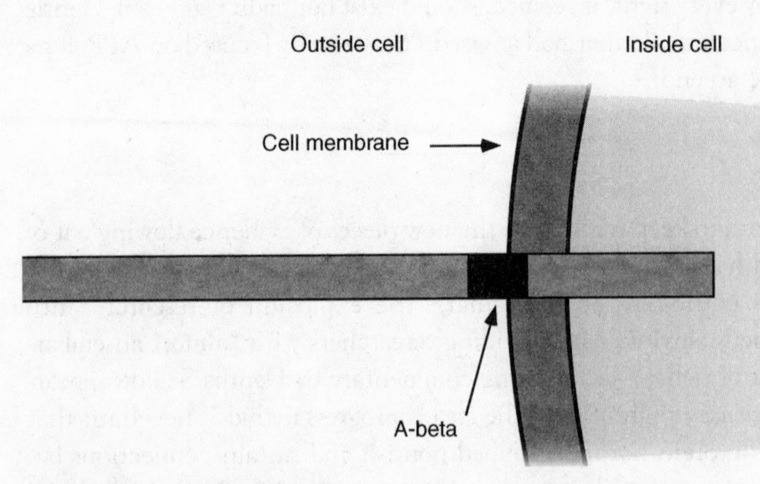

FIGURE 6.1 APP protein and its A-beta snippet.
Illustration: Robert D. Moir

about its extracellular business. The sixty-four-thousand-dollar question, then, was how did only the short strip of A-beta wind up in the middle of amyloid plaques? Did a healthy cell actively secrete A-beta through some elaborate clipping mechanism by specific proteases, whereupon A-beta aggregated into amyloid? Or did a dying cell shed the entire APP protein, whereupon it formed plaques, then got trimmed down by other proteases to leave only A-beta in the plaque's core? If there was a prayer that drugs could reduce the production of amyloid, they couldn't be developed until the veritable road to amyloid was defined.

Speculation ran wild about where, exactly, an enzyme cut APP's outer large portion. Numerous investigators were on the prowl. In mid-1990, in reports spaced a few months apart, molecular biologist Sangram Sisodia at Johns Hopkins School of Medicine and biochemist Fred Esch at Athena Neurosciences finally delivered. Sisodia demonstrated where the cut in all probability happened, and Esch determined the actual cleavage site. They divined that as a matter of course a protease made a cut right in the *middle* of A-beta, setting free the protein's long exterior from that point outward. This represented a completely normal activity, a routine event for the APP protein. And look what it meant! Because this clip occurs in the middle of A-beta, A-beta gets broken in two—deactivated! trounced!—making it impossible for it to aggregate. This normal, "good" pathway didn't lead to amyloid.

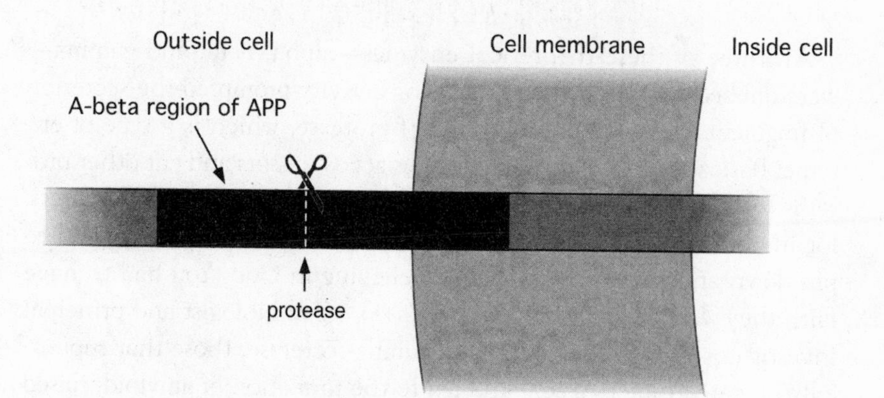

FIGURE 6.2 APP protein cleaved by "good" protease.
Illustration: Robert D. Moir

The unseen, unknown, and therefore hypothetical protease that made this "good" cut was given the name *alpha*. But the question still nagged: How did just the A-beta piece get free and wind up *intact* in plaques? A logical conclusion followed: Possibly two other proteases snipped A-beta at each end, detaching it from APP and sending it on its aggregating way. These unfound, unseen, hypothetical proteases were called *beta* and *gamma*. Most importantly, if they really existed, they had to set in motion an abnormal "bad" pathway that led to amyloid.

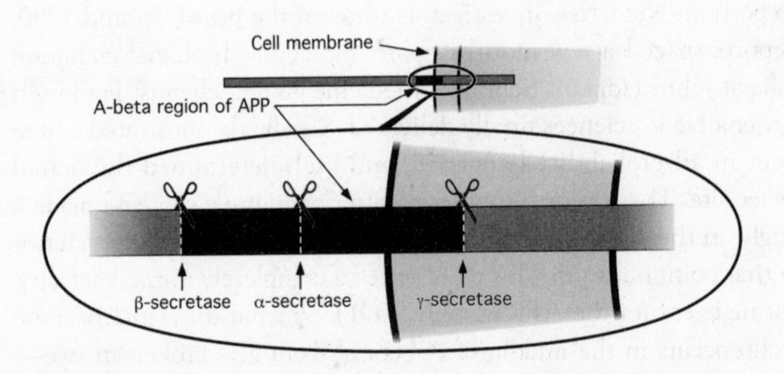

FIGURE 6.3 APP protein and the sites of the three secretases.
Illustration: Robert D. Moir

All three of these theoretical enzymes—alpha, beta, and gamma—were dubbed *secretases;* their catalyzing activity prompted the secretion of fragments. (A secretase is a type of protease, which is a type of enzyme. Basically all three type molecules act as scissors and cut other proteins.) What better term for molecules that were such deep, dark secrets, for in coming years the secretases would prove exasperatingly hard to pin down and identify. "It was like believing in God. You had to have faith they were there," says Wilma Wasco, a cell biologist and principal investigator in my unit. Beta- and gamma-secretase, those that supposedly liberated intact A-beta and led to the formation of amyloid, raised exciting prospects. If either one could be caught and chemically deciphered, a wrench might be thrown into its works, keeping A-beta from being cut loose and clumping into plaques. Bench detectives immediately set out in search of the secretases. From that point on, one regularly heard rumors that beta or gamma had been found, without strong confirmation to follow.

I first crossed paths with Sam Sisodia—who sported an impressive Groucho moustache and was originally from Udaipur, India—at the 1990 Dahlem Conference on neurological disorders in Berlin. I'd seen his name in many a journal report—as he had mine—which suggested to me he was a senior gray-haired sort. We took one look at each other, and had the same response. "I thought you were a whole lot older!" he exclaimed. "Gee, I thought *you* were!" I said. We had a big laugh over this and became friends on the spot.

All this while investigators had continued to scour chromosome 21 for the "other" gene. Three years had passed since the hunt had started. The DNAs of scores of Alzheimer families had been probed without a fleck of hard data gained as to the gene's whereabouts. Many researchers were on the verge of concluding, if they hadn't already, that the object of their fruitless search indeed must be a phantom. Besides, faint signs of Alzheimer-associated flaws on other chromosomes were gradually stealing into the picture. Genetic forms of cancer, heart disease, and other major illnesses were being tied to multiple genes, and acceptance was growing that inherited Alzheimer's might similarly result from several faulty genes on multiple chromosomes and that the field should broaden its search.

Still, it was tremendously hard to give up on chromosome 21, especially given that things at the molecular level can be so incredibly intractable and hard to measure that scientists can be duped into thinking that they don't exist, when in fact they really do. We scientists constantly have to remind ourselves of a commonly used adage: *The absence of proof of something's existence isn't necessarily proof of its absence.* Just think of it this way. If you look for a wallaby in a grassy plain and don't find one, it doesn't mean that no wallabies are present. They may be too far away to see; or they may be hidden behind a knoll, rolling on their backs; or they may be standing in a streambed, having a good long drink. *The absence of proof is not the proof of absence.* My lab crew hears me say this time and again. Despite all the futile hunting high and low on chromosome 21, until all possibility was exhausted it remained plausible that it was home to an Alzheimer flaw.

The day I finished typing up my thesis, appropriately enough, was the running of the Boston Marathon—Patriots' Day of April 1990. Hunkered over the keyboard, I wasn't entirely sure I was going to make it to the finish line by hand-in time. Later that day, I tuned into the Red Sox and wished I hadn't. They were being cremated by Milwaukee, the score ending up 18–0. I was afraid this might be a bad omen for my thesis. In early May, however, I delivered my two-hour thesis defense in Building B in the Med School quad and all went smoothly, although no department party happened afterwards, as traditionally took place. The one faculty member who appeared was Huntington Potter, bearing a bottle of Veuve Clicquot with "Congratulations Doctor Rudy" written across the label. Only a diploma was needed before I could leave behind my Ph.D. travails and sail into the world—a free neurobiologist. Happily, Gusella's molecular neurogenetics unit would remain my home for the foreseeable future. Just as Gusella had done after receiving his doctorate from MIT, I would bypass postdoctoral training and join Harvard's junior faculty as an instructor of neurology.

But when commencement day arrived, it brought one final bump. Just before the ceremonious march into Harvard Yard, following instructions—perhaps better than I had for five years—I went to my assigned line only to find myself surrounded by a sea of unfamiliar faces. My consternation grew when I saw Guy Rouleau's fox-red beard and then the familiar faces of other classmates bobbing way across the yard. I soon learned I was amidst Divinity School graduates! Was this some sort of practical joke? It was pointed out that the crow's-foot emblem on my robe was scarlet, for Divinity School, and not green for Division of Medical Sciences. All I can think is that when I rented my gown at the Harvard Coop, someone must have misread "DMS" for "DS" and given me the wrong robe, which somehow led to my joining the wrong line. It didn't take me long to race across the yard and join my classmates. I'd make a lousy priest.

Right before graduation, I had received a call from Blas Frangione, a biochemist and authority on amyloid at New York University's School of Medicine. In 1982, a year before Glenner isolated Alzheimer's A-beta peptide, Frangione—who happened to be a distant cousin of Maria Frangione, Jim Gusella's wife—had distinguished himself by isolating the very first amyloid protein associated with a brain disorder: an inherited amyloidosis seen in Icelandic families, the symptoms of which begin with small strokes that can lead to dementia. Its amyloid was restricted to the brain's blood vessels, and its protein was different from A-beta. Frangione told me he was working with the DNA of a Dutch family who exhibited yet another example of an inherited amyloidosis seen in the brain's blood vessels. Could I help him sequence the family's DNA? I said I'd be glad to. He left me with a small pearl, although I didn't think much about it at the time. While the Dutch disease's symptoms didn't resemble Alzheimer's, its amyloid *was* made up of A-beta. And he thought that those afflicted possibly could have a mutation in their APP gene.

Julia and her family had been told that in all likelihood her disorder was Alzheimer's, not schizophrenia, and that she might live for another five to eight years. Eleven years later, in 1978 and at age fifty-four, Julia Tatro Noonan escaped this world. Her family had struggled to provide her with good care, constantly shifting her back and forth between home and outside facilities. Home care and its round-the-clock aides were exorbitantly expensive, draining nearly every cent of John Noonan's income. And when she was at home, Julia would cry so, say such strange things, and be so anxious and disoriented that whatever family life John and those at home had scraped together would unravel. Aides would depart for one reason or another, and Julia would be moved to yet another nursing home, but never for long. Many establishments didn't want the responsibility of a patient who was apt to wander all hours of the night and be disruptive. And so Julia would come home again. Before she died, John Noonan made the hard decision of divorcing Julia. Without means, she had become a ward of the state and therefore eligible for caretaking funds through Medicaid.

— *seven* —

Mutations, Revelations

Then more
and more,
forming a
foreshortened
corridor or
niche of yeses . . .
—Kay Ryan, "Yeses"

Many of us felt thoroughly baffled as the new decade bore down on us. Why had no Alzheimer-related DNA mutations turned up in the scores of early-onset families being analyzed on various continents—especially in the APP-amyloid gene? The disease's early variety was so obviously driven by an inherited genetic fault. But only by identifying a mutation could we hope to demonstrate proof of a direct cause-and-effect between a defective gene and the disease.

Seated on the terrace of the Hotel Villa Del Mare on a warm July evening in 1989, high above the Gulf of Policastro in southwestern Italy, Guy Salvesen and I swirled down grappa and stared out over the jet-black water and fishing boats moored far below, all the while exchanging nuggets about the state of amyloid research. Salvesen was a biochemist then at Duke University, and both of us had landed in Maratea for a NATO workshop on brain-related protease inhibitors. After swapping news and gossip from other benches, we settled into a debate over the questionable existence of an Alzheimer mutation on chromosome 21. Even though most in the field were beginning to have their doubts about the "other" gene and shift their search to other chromosomes, Mass General's earlier reported linkage to chromosome 21 still made one wonder.

"So what do you think," mused Salvesen. "You think there's a defect on 21?"

"Tell you what. If there really is, then we'll see a shooting star—right there," pointing my finger toward the star-studded heavens. "Right now." It was a game my twin sister Anne and I had often played as kids. No sooner were the words out of my mouth than a fantastic blazing streak shot down directly in front of us. We sat stunned and then continued quite late into the night with the grappa. I wondered for days if this sign meant anything beyond simple coincidence.

That same July, hundreds of miles north at the University of Antwerp, a molecular geneticist was still searching for an Alzheimer mutation, and not only on chromosome 21, but in its APP gene. A straight-shooter, Christine Van Broeckhoven didn't shy away from speaking out about the tendency of medical journals to publish sensationalist "rubbish" over substance or, for that matter, the male-dominated field's too-few meeting invitations to women scientists. In the same vein, she didn't shirk from following her instincts—even as pertained to a gene that had been shaken inside-out for an error.

Motivating Van Broeckhoven was a one-of-a-kind opportunity. Just above Belgium, living in two coastal towns along the Netherlands' flat coast, were four distantly related families who were afflicted by an extremely rare inherited amyloid disease in which amyloid damages the brain's blood vessels, accounting for brain hemorrhages that usually prove fatal by age fifty or sixty. The disorder, which is the very same one that Blas Frangione was exploring, goes by the long-winded name of *hereditary cerebral hemorrhage with amyloidosis of Dutch type*, or HCHWA-D. Unlike Alzheimer's, HCHWA-D doesn't overwhelm the cerebral cortex with plaques and tangles, and prior to the onset of stroke, few of its victims experience dementia. Yet, as mentioned, the amyloid deposited by this rare condition in the brain's vasculature is formed by the A-beta peptide, just as in Alzheimer's. This insight had been obtained by Blas Frangione's lab two years after Glenner had isolated A-beta.

Van Broeckhoven had a strong hunch she'd find a mutation tied to the Dutch disease in the APP-amyloid gene, and very conceivably, she reasoned, where the Dutch defect lurked an Alzheimer defect might lurk. Having access to DNA from blood samples of living HCHWA-D family members as well as from the brain tissue of those deceased, which had been collected by pathologists at the University of Leiden in the Netherlands, her lab in conjunction with Leiden researchers cranked up their investigation. By the winter of '89/'90, linkage had been established. Yes! The Dutch disorder was associated with something amiss on chromosome 21. More analysis was required to tell if the defect actually sat in the suspicious APP gene.

Playing a peripheral role in the research was Van Broeckhoven's close friend and collaborator, John Hardy, a biochemist who headed a lab in the Department of Molecular Genetics at St. Mary's Hospital Medical School in London. I'd increasingly gotten to know and like Hardy—a somewhat disheveled Paul McCartney look-alike with longish, dirty-blond hair. He had an admirable way of cutting-to-the-chase and distinguishing meaningful data from slop. Like many of us, Hardy had lost faith in finding an Alzheimer flaw in the APP gene, but when Van Broeckhoven began closing in on the Dutch-disorder mutation, she gave her friend a nudge, telling him he really should take another look at the APP gene. The two scientists shared a pool of Alzheimer families, one of which—a family Hardy had brought to the research—actually had been signalling an error on chromosome 21. Hardy's lab therefore promptly renewed its sequencing of APP under the direction of molecular geneticist Alison Goate.

When Van Broeckhoven submitted her account to *Science* of the Dutch disorder's link to chromosome 21, she and her labmates in actuality were quietly celebrating a far grander coup. They'd isolated the disorder's very mutation, recounts Van Broeckhoven, and it sat right in the APP gene, just as she'd predicted. She had decided to wait and publish this much bigger news in a separate paper. But then, to her astonishment, she heard that another group—Blas Frangione's New York University lab—also had dredged up the Dutch mutation and that their report would run in the same June 1990 issue of *Science* that Van Broeckhoven's chromosome report was slated for. This, by the way, represented the sequencing work Frangione had discussed with me just before I'd graduated, but in the end had assigned to his post-doc Efrat Levy, the resulting paper's lead author. Van Broeckhoven had had no idea that Leiden researchers had been sharing the Dutch families' tissues with Frangione's team.

The New York team's narrow edging out of Van Broeckhoven's team thus ended a very close bicycle race of sorts. As John Hardy observes about scientific endeavors, so often it happens that two teams are pedalling neck and neck without necessarily even being aware of it—inching forward, falling back, but basically so cheek by jowl, it's as though their bicycles were attached. Then, in the race's final moments, one cyclist pulls ahead. "It can happen very quickly in science," observes Hardy. "In a matter of weeks, one paper can achieve stardom, while another paper gets scuttled."

Van Broeckhoven's linkage paper nonetheless provided important backup confirmation of Frangione's mutation paper. But more significantly, due to her earlier tip to John Hardy, his London lab was already busily reanalyzing the APP gene in its Alzheimer family of interest. When word broke that Frangione had speared the Dutch mutation, Hardy, Goate, and coworkers found themselves in an ideal position. "We went straight to the same place on chromosome 21 in our Alzheimer's family," recounts Hardy. Within a week they spied what hundreds of researchers in dozens of labs had been persistently pursuing for so many long years: a tiny Alzheimer-causing substitution of one base for another that lay roughly seventy bases away from the Dutch mutation in the APP gene! The base T where normally lies C. In the gene's encoded protein, an isoleucine amino acid where normally lies a valine. Human normalcy is that easily ruined!

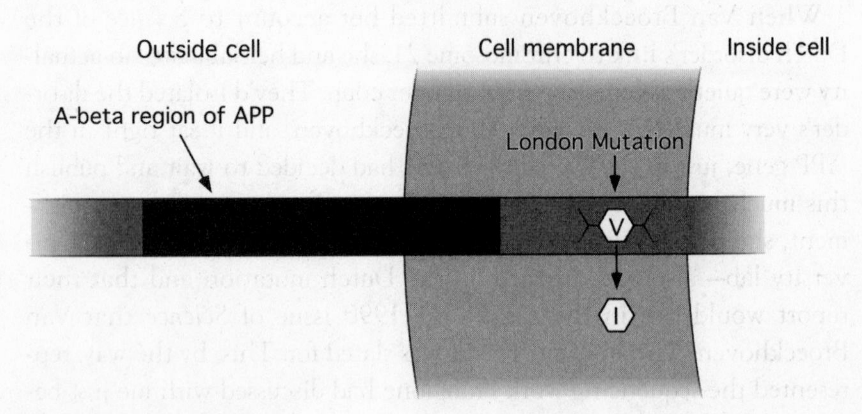

FIGURE 7.1 APP protein: site of the first-found Alzheimer's mutation. Illustration: Robert D. Moir

Hardy's lab went bonkers. For the first time ever a cause for Alzheimer's disease had been identified. In many respects, Blas Frangione—first by observing that the Dutch and Alzheimer's amyloid was composed of a similar protein fragment, then by his early location of the Dutch mutation—both laid the groundwork for this momentous breakthrough and was the canary that led the mouse to the cheese.

As incredibly sublime as the discovery was, there was a disconcerting twist. None of the DNAs of the other twenty-two Alzheimer families kept by Hardy's lab carried the mutation. Writing up the one-family mutation felt a bit flimsy, but Hardy was on the verge of doing so when he suddenly remembered that stored in the lab's freezer was the DNA of an Alzheimer family that he and Alison Goate had received two years earlier from Allen Roses at Duke University. Perhaps its cell lines carried the same defect. "Allen didn't believe there was an Alzheimer's defect on chromosome 21," recalls Hardy, "and because we did, he had passed his family along with a certain element of bravado." At the time, Allen Roses's opinion that neither A-beta nor amyloid played an integral role in the disease was on its way to becoming famously well known in the field.

"Once we remembered the other family, we screened it, and it had the same APP mutation we'd seen in the first family," recounts Hardy. "That was a very important piece of information, really, because finding the mutation in two unrelated families proved, in my mind, that it was a disease-causing pathogenic mutation. The final proof, then, was because of that family." Yet Hardy soon found himself "caught in a bit of a conundrum," he says, for he couldn't tell Roses about the finding. "We went to patent the mutation through Imperial College"—the overseer of St. Mary's Hospital—"and lawyers there told us that under no circumstances should we tell anyone outside the lab what we had found"—until their paper was in print—"because the patent would be valuable and we would be undermining it."

Hardy followed these instructions, and, he says, sorely regrets doing so. After the journal *Nature* accepted the Hardy/Goate report of APP's mutation, one of the paper's reviewers leaked the information, according to Hardy, and the leak, he says, trickled into the ear of a *Newsweek* reporter, who phoned Allen Roses for comments. (Several years later Roses recounted this differently, telling *Science* that someone had anonymously faxed him a copy of the *Nature* paper—from London's Paddington Station.) The combination of hearing about the APP mutation from an inappropriate source and learning that he wasn't included among the paper's authors made Roses go "completely ballistic," says Hardy. Hardy, who had felt no compunction to include Roses as an author, since Roses had signed off on chromosome 21, notes that he actually ended up sending Roses a letter about the finding, but that it arrived after the damage was done.

Within no time, Hardy received a volley of hard-hitting phone calls. A fuming Roses was the first to call; then Bob Williamson, Hardy's boss and the chairman of St. Mary's molecular genetics department, who'd been called by Roses, according to Hardy; then the NIH, who backed Roses's work; then lawyers from Duke, Roses's university. When under assault, Hardy acknowledges he isn't exactly a shrinking violet. "I react very badly to being shouted and screamed at." Adding to the heat was "the thought that this first-found mutation might be *it*—all of Alzheimer's might be sorted out. And here the mutation had been found by a small English group." Attempting to beat down the flames, Hardy included Roses and Margaret Pericak-Vance, Roses's lab partner, as co-authors on the paper, but it didn't placate matters. "Allen and I have never gotten along properly since," says Hardy. "I think the story shows that I was in the wrong. I look back and I wouldn't have done now what I did then. I would have just ignored the patent lawyer and informed Allen."

Months before the Hardy team's *Nature* write-up reached ink in February 1991, whisperings about the Londoners' trapped mutation swirled through the field. The news reached me via one colleague or another not long after those of us who made up Jim Gusella's neurogenetics entourage had moved into Building 149, the colossal 650,000-square-foot former warehouse in the Charlestown Navy Yard that Mass General had acquired and refurbished for the housing of a considerable chunk of its medical research, among other uses. Rumor had it that during World War II, the world's single largest collection of torpedoes had been stored there. Now that the cold war was over and the war on disease was rapidly escalating, it seemed ever so appropriate that Building 149 instead was stockpiling DNA.

So the shooting star over the Gulf of Policastro hadn't lied! Chromosome 21 did hold an Alzheimer flaw—smack in the APP gene! After years of waiting and watching mutations tied to other diseases exposed at a quickening pace, our field finally had isolated its own denizen. But we were left with burning questions: Just how much of inherited Alzheimer's was caused by the mutation on APP? Had the rest of us somehow missed this tiny, single-base error when scanning the DNA of our own early-onset families?

Throughout the field, the APP gene in hundreds of kinships would have to be reexamined. There was fierce hope that the "London mutation," as it came to be called, or other mutations in APP would explain the disease's descent in many of the collected early-onset families. But because the gene and its chromosome had been so arduously examined in so many families, with little or no incriminating data revealed, there was equally fierce skepticism. Mass General's four families, for instance, had been so thoroughly analyzed, I doubted the presence of a London mutation in these cases.

Raising further doubts, new clues were implicating other chromosomes. At the University of Washington, neurogeneticist Gerard Schellenberg and coworkers had analyzed a group of its early-onset Alzheimer families—a kinship known as the Volga Germans—and had come away with telltale evidence of an impaired gene they were fairly certain wasn't on chromosome 21, which cancelled out APP. Peggy Pericak-Vance in Allen Roses's Duke lab, meanwhile, had indicated that she and coworkers had tracked the disease to chromosome 19 in a few small late-onset families. If this line of inquiry held ground, it would be of unprecedented value. Epidemiological studies were suggesting that an inherited feature sometimes contributed to the most common variety of Alzheimer's—its appearance over the age of sixty—and here was the first DNA sign of such.

When, in February 1991, the Hardy team's London mutation at last was officially disclosed in *Nature*, an unnatural silence hung over its authors' London lab. "The phone lines went dead. Absolutely nobody called," recalls Hardy. "It was weird. The only person who contacted me was Ivan Lieberburg from Athena Neurosciences," a biotech company that was interested in purchasing the mutation's patent from Imperial

College, and did so within a month. "To us in the lab it felt like people were sheepish, because what we had done was so simple." They'd gotten a good tip, gone to the cell lines of their one Alzheimer family that hinted at a disturbance on chromosome 21, and there lay the mutation. The silence of the labs probably also stemmed from the fact that so many of us already knew via the grapevine about their discovery. The media, on the other hand, was anything but hushed. "Gene Mutation That Causes Alzheimer's Is Found," roared out the *New York Times* on its front page. "Family Links Offer Hope of Alzheimer's Disease Cure," the London *Times* optimistically headlined.

Seven or so years had passed since George Glenner and Cai'ne Wong had isolated the A-beta peptide from the disease's cerebral plaques, and all at once their exploit was bearing serious fruit. The field had hold of the scintillating theory of the secretases—chemical scissors that might be responsible for excising A-beta fragments that lumped into amyloid. And now we knew that at least two, and maybe more, early-onset families had an APP flaw. Did that mutation affect how the secretases interacted with APP and freed A-beta? Around the time the London mutation was being brought to the surface, yet another APP finding rocked the field, guaranteeing to lure all the more researchers into amyloid's web.

At a September 1990 meeting in Tokyo, hundreds of predominantly Japanese scientists listened intently as Bruce Yankner, a neurologist, told how he and his coworkers at Boston's Children's Hospital had added the excised A-beta fragment, which they'd synthesized in the lab, to cultures of neurons from the hippocampus of rats. (The hippocampus, you'll recall, is one of the brain regions in humans so severely affected by Alzheimer's.) Within two to three days, approximately three-quarters of the rats' brain cells had died due to what appeared to be A-beta's overwhelming toxicity. This amounted to one of the field's hitherto most rousing revelations. On and off since Alois Alzheimer's day, scientists had puzzled over whether some lethal aspect of plaque amyloid caused the disease's degeneration of neurons. Yes, too much of the microscopic gook inundated the brain, although too much of something doesn't always mean it's bad. An earlier finding by Yankner and Rachael Neve had implied that some part of the APP protein was neurotoxic. But here, apparently, was a first-time demonstration that APP's A-beta peptide, once cut loose, might be the disease's actual assassin of neurons and their synapses.

For Yankner, his delivery of this major insight proved fairly agonizing. "I almost didn't make it through the talk," he recalls. Relatively new to our research field and new to public speaking, he wasn't only nervous, but also considerably rattled by the Japanese love of photography. Flashbulbs kept exploding in his face, causing him to lose his train of thought.

Two months later at the annual Neuroscience meeting, Yankner reported that his team had gone a step further. I was in attendance and, like everyone else in the room, all ears. This time Yankner and his teammates had injected A-beta directly into rat brains, and here again a substantial toxic effect had ensued. After describing this outcome, on the spur of the moment Yankner salted his talk with a few words he remembered from somewhere. It had "not escaped" his team's notice that inhibiting amyloid's neurotoxicity might serve as a therapeutic approach to Alzheimer's disease. His word choice harkened back to Watson and Crick's 1953 historic paper on their discovery of DNA's structure: "It has not escaped our notice that the specific pairing we have postulated immediately suggests a possible copying mechanism for the genetic material." Recounts Yankner about his rendition, "I said it as a joke, and the humor succeeded."

While the evidence that A-beta fragments might spell death for neurons wasn't of the same magnitude as Watson and Crick's disclosure, if Yankner's finding was for real, it would be no joke. It might provide knowledge that could help to eventually prevent Alzheimer's assault on neurons. Yankner's description electrified the field, but along with the great excitement, there was skepticism too. His work wrapped around the basic assumption that A-beta free-floated for a spell in the brain. Since A-beta only had been found stuck in the plaques, no one had actually witnessed it free-floating, which suggested that if it did so, it must exist at very low levels. Yet the levels of A-beta used by Yankner were quite high—high enough to force toxicity, perhaps. Something else cast doubt: If A-beta was toxic, wouldn't the same be true of the plaques? Yet these clusters sometimes sat right beside seemingly healthy neurons. Conspicuously, too, brain regions filled with plaques didn't always correlate closely with those exhibiting extensive neuronal death.

Could it be that once A-beta consolidated into plaques, it stopped being toxic? This fit with a theory that I and other researchers were exploring: If A-beta existed in free-floating form in the brain and was lethal, maybe the plaques actually worked as a protective mechanism that took A-beta out of solution so it wouldn't harm the body.

Since I was in the throes of analyzing how the APP gene was regulated in the brain, I resisted the temptation of going to the bench to try out Yankner's experiment. But several other research groups ran the test. Some saw what Yankner's team saw—a marked deterioration of cells in the rodent hippocampus commensurate with the amount of A-beta they injected. Yet others, even after repeat attempts, and at the expense of many rats, couldn't replicate Yankner's results. The more time they spent trying, the more exasperated they became, and the more exasperated they became, the more they blamed Yankner for either not disclosing the finer details in his protocol or for overstating his results.

As would come to light, there was a very real reason why some groups could and others could not replicate the experiments, an explanation that Yankner says he wasn't aware of initially. In the test tube A-beta peptides gradually self-associate and form fibrils before clumping into plaques—and *it was the fibrils that appeared to be toxic*! Christian Pike and Carl Cotman at the University of California, Irvine, provided this gift of an insight. If the peptides weren't prepared properly or didn't have time to age, the results didn't yield the same toxic effect. Yankner's experiments happened to have tapped into this intriguing prerequisite for toxicity.

All told, the controversy surrounding Yankner's work reflected how crucial it was to correctly gauge whether A-beta was potent enough to kill neurons. Yankner's investigative thread had the positive effect of opening a Pandora's box of tough questions that sorely needed to be examined—and continue to be. For although A-beta's toxicity has been proven under certain conditions, to this day the jury remains undecided as to whether this toxicity is capable of killing neurons to the

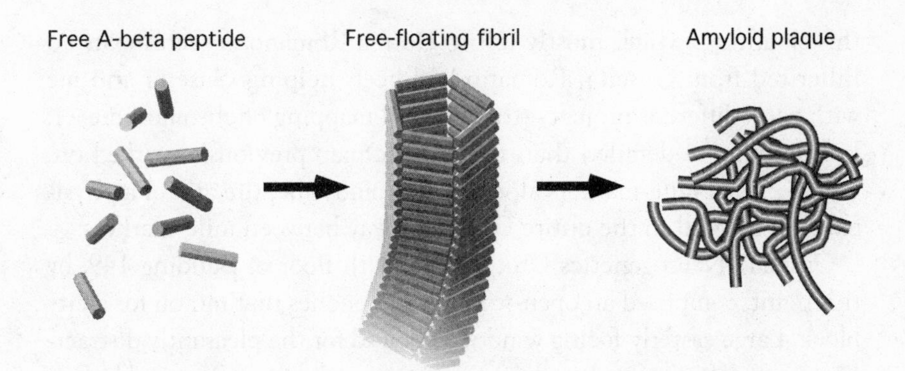

FIGURE 7.2 The A-beta peptide and its aggregation first to fibril then to plaque. Illustration: Robert D. Moir

degree seen in Alzheimer's and therefore represents the disease's primary insult. "That A-beta's toxicity is the cause of Alzheimer's has not been proven. I would be the last one to dispute that," says Yankner today. "But right now, there is no other mechanism so strongly supported by data that can explain the degeneration of neurons in Alzheimer's."

From the time I finished my doctorate up to the published capture of APP's London mutation, things in my niche in Jim Gusella's unit had been dishwater-dull. But once APP's defect was officially on the table in early 1991, goings-on in my lab immediately began heating up. (A few months before, I'd been given my own lab, complete with two entire benches.) It was necessary to haul out the DNAs of Mass General's Alzheimer families and again examine their APP gene for a mutation, this time through raw sequencing. Possibly our previous genetic-linkage tests ruling out APP had lied. Moreover, new Alzheimer kinships had been added to the MGH's growing stable and just might harbor the defect. In time, my lab received its first large federal grant to cover this sleuthing, which mostly fell to Donna Romano, a technician I'd inherited from Gusella. Romano had been helping Gusella and me with yet a different project—the physical mapping of chromosome 21. Infinitely more detailed than the genetic map previously worked on, which placed mile-markers along the chromosome, the aim of a physical map is to fill in the entire DNA roadway between mile-markers.

Gusella's Neurogenetics Unit on the sixth floor of Building 149, by this point, comprised an open-wall row of benches that ran on for a city block. Large easterly-facing windows allowed for the pleasantly distracting view of barges and tankers nosing in and out of Boston Harbor. Within the unit, the focus on Alzheimer genetics was but one piece of the pie. The Huntington gene's exact error on chromosome 4 proving far more elusive than its home chromosome had been, that grueling search was still in progress. Dozens more of Gusella's battalion were investigating genetic abnormalities tied to dystonia, Batten's disease, Lou Gehrig's disease, Wilson's disease, neurofibromatosis, and other neurologic sickness.

The previous summer, President George Bush had signed a proclamation designating the 1990s "The Decade of the Brain." There was so much more to learn about this "most magnificent—and mysterious" wonder of creation, especially the illnesses that stole from it, stated the proclamation. For those of us under Gusella's roof working at the forefront of this frontier, how large the opportunities, how special our given chance.

I was well aware that at least a dozen other labs were checking their Alzheimer cell lines for defects on the APP gene. As I told Romano, she'd have to come up with a faster method of scanning our family DNAs, which amounted to over 200 cell lines, or we'd fall behind. Much to her credit, Romano did so, by utilizing a revolutionary new tool called PCR, or polymerase chain reaction, in combination with her own modified sequencing technique.

To DNA scientists, PCR was fast becoming invaluable. It allowed a scientist to take a stitch of DNA obtained from blood, bone, or other tissue—or DNA from a person's established cell line—and amplify it millions of times within just a few hours, rapidly providing enough DNA for sequencing or other work. Gone were the laborious steps of manually cutting DNA, inserting it into a vector, and propagating it in bacteria. "With PCR, working up and scanning ten people's DNA took me only two days, instead of two weeks," notes Romano. "It accelerated the work so phenomenally that it seemed as though, before PCR, I'd been working in slow motion."

Developed in the mid-'80s by Kary Mullis at Cetus Corporation, PCR harnesses a unique enzyme from bacteria that live in hot springs and deep-ocean thermal vents. Taq polymerase, as the enzyme is called, can read single-stranded DNA and replicate it, at the same time withstanding the boiling-hot temperatures periodically called for in DNA's amplification process. PCR represented such a technical leap forward, it would earn Mullis a Nobel Prize as soon as 1993. Bone from a 30,000-year-old Siberian horse; a drop of dried blood found in a white Ford bronco; dried saliva on a licked envelope. Wherever an organism has left behind a cell, there's the potential of using PCR's wizardry to recover DNA. Since DNA is a chemical, it crystallizes when it's dried out, but mixing it in water can bring it back into solution—similar, say, to sugar crystals—whereupon PCR's enzymes can amplify it. DNA older than 100,000 years, however, may be too deteriorated for recovery, scientists are realizing. So the *Jurassic Park* fiction of cloning dinosaur DNA to resurrect those long-ago creatures—even dino DNA found in a blood-sucking insect preserved in amber—remains only fiction, at least for now.

By the summer of '91 Romano had sequenced the APP gene from patients in thirty-odd Alzheimer families and twenty-six individual cases. No London mutation had turned up. The mutation's known vicinity in APP just might signify a hot spot where different Alzheimer mutations lurked in our families, so the logical next step was to explore adjacent areas in APP. "It was exciting to have a quick way of scanning DNA that wasn't being used in other labs and realize that I might come across a mutation no one else had seen," Romano recalls. A weekend scuba diver, she compares mutation-hunting to the novelty of encountering the unexpected underwater. "Scanning for a mutation—well, it's something like coming across a nudibranch, a slug-like mollusk that's apt to be very elusive and all the more beautiful when you spot one."

But the year lengthened, and still Romano turned up nothing. Then in October, word reached us: A second Alzheimer mutation in the APP gene had been found by Merrill Benson's Indiana University lab in one of its families. It changed the very same amino acid in APP's protein as the London mutation did, but involved a different DNA substitution: an A where normally lies a C. Hard on the heels of that news came still another report from John Hardy's group. They had detected a third APP mutation in the very same area. Frustrated, I'm afraid I heaped pressure on Romano. "Tremendous pressure," acknowledges Romano. "In scuba-diving terms, it was more than a couple of atmospheres of pressure." How come these other teams were finding mutations, and we weren't? Get cracking! I told her. At conferences I was getting a lot of collegial flack for not having found an APP mutation, in particular from John Hardy, who was starting an E-mail club expressly devoted to APP mutations. At one point, Jerry Schellenberg, who hadn't turned up any mutations either, called me to commiserate. "I'm really getting paranoid," he said. "Are we doing something fundamentally wrong?"

There was an easy answer to why our labs were coming up empty. Toward the end of 1991 we would conclude that the MGH's Alzheimer families had no APP mutations. Nor did Schellenberg's families, nor most of those kinships gathered by other groups. In all these cases the disease had to originate from a different inherited mutation or mutations elsewhere in the genome. As one after another team rolled out their data, it became clear that mutations in the APP gene accounted for an extremely nominal percentage of inherited Alzheimer's. To date, the eight isolated explain an estimated 2 to 3 percent of inherited early-onset cases, or far less than 1 percent of all Alzheimer cases, young and old.

What about Mass General's earlier report of a link between Alzheimer's and some anomaly on chromosome 21 seen primarily in its Italian family? The claim that had kept so many in the field combing chromosome 21 for the "other" gene? Well, the lod scores underlying that data had fibbed, as statistical measurements sometimes do. The paper was dead wrong. As the future would bear out, the DNA flaw in the Italian family as well as MGH's other three original Alzheimer families resided on an entirely different chromosome. What an outstanding irony we'd be left with! Because of Hyslop's finding that hinted at something amiss on chromosome 21, John Hardy's lab had determined it too had a family linked to chromosome 21. Yet while the original MGH data turned out to be false, the Hardy team's linkage to chromosome 21 in one family had turned out to be real! Moreover the culprit on 21 was the originally suspected APP gene!

"It was a bizarre conundrum that the original linkage report from Mass General was wrong," recounts John Hardy. "Bizarre, because it was wrong in the right place."

The location of the first-found APP mutations exhibited, perhaps, the very essence of their guilt. In APP's protein, they sat only a few amino acids away from where APP was presumably cleaved by one of the secretases—gamma—to generate the A-beta fragment. The implication was at once obvious and tantalizing: It would seem the mutations somehow increased the likelihood of the secretases' cuts and therefore the release of abnormal amounts of A-beta. Should this be true, isolating the secretases or deducing their activity could very well hasten the development

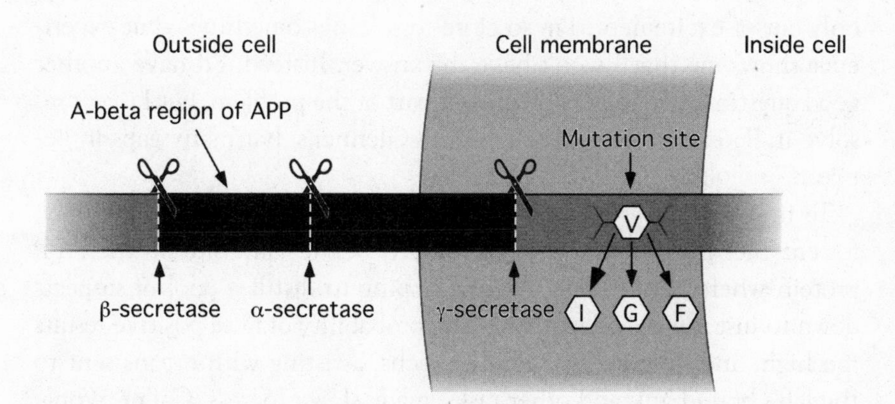

FIGURE 7.3 APP protein: the first three discovered mutations and the clipping sites of the three secretases. Illustration: Robert D. Moir

of drugs that could target and block the activity of the secretases, impeding amyloid's cerebral buildup.

Burning the oil late into the night in his UCSD lab, George Glenner was one of the many researchers who had taken on the challenge of hunting down beta—the other secretase that freed A-beta. Beta seemed an easier secretase to chase down than gamma because unlike gamma it didn't cleave in the membrane. That gamma did so was perplexing, since the membrane was uncharacteristic territory for a protease. Glenner and Joy were more dedicated than ever to softening Alzheimer's blow for patients and their families, and they had opened a second Alzheimer day-care center, which, like the first, was filled with as many patients as the staff could manage and the rooms could hold. Yet ever since Glenner's cornering of the A-beta fragment, and the research community's subsequent surge forward into the genetic gut of the disease, Glenner had felt a special exuberance toward the bench, and the dream of obtaining the final answer to Alzheimer's seemed so close, he dared hope to see the disorder solved within his lifetime. Perhaps he could be the one to completely solve it. Yet he was wise enough to realize, however, that after making one major contribution, oftentimes a scientist has no more gold to spend.

"I love being a scientist, because every day I am posed with new questions," said Henry Wisniewski, Glenner's longtime friend and colleague, speaking of a passion that similarly drove Glenner. "Often in the puzzle, one piece isn't fitting and ruining the rest of my solutions. I lose sleep, only out of excitement. I'm so close to solving something—but experience shows me that I won't have the answer. Instead, I'll have another good question, and maybe I'll answer part of the problem, but I won't resolve it. There are breakthroughs like Glenner's, but many gaps in between."

To find the beta-secretase, Glenner and his technicians would look for enzymes in the brain that cut exactly at the same site on the APP protein where beta did its cutting, hoping to distill a pool of suspects down to just the enigmatic beta. The probability of false positive results ran high, and Glenner's handful of techs, assisting with organs sent to the lab's brain bank and other tasks, made slow progress. Cai'ne Wong, meanwhile, had moved on. Eventually he would join the Rocky Mountain Laboratories in Hamilton, Montana, where today he researches prion diseases.

UCSD was well on track to becoming one of the foremost educational centers for neuroscience in the country. Now, in 1991, the university had the enviable Alzheimer's research triumvirate of Glenner, Robert Katzman, and Bob Terry. The latter two scientists had arrived from Albert Einstein in the mid-'80s when Katzman was invited to head UCSD's neuroscience department, which had convinced Terry to make the move west as well. Their disciplines being so compatible—Terry the neuropathologist and Katzman the neurologist—they made for an indefatigable pair in their efforts toward Alzheimer's, both at the scientific and administrative levels. By the early '90s, both gentlemen were widely respected doyens in the field, having fathered and grandfathered countless scientists. Yet the Glenner-Katzman-Terry triumvirate, alas, never actually worked as such. Even though they existed side-by-side—with Glenner in the Basic Sciences building, and Terry and Katzman next door in the Medical Teaching Facility—as everyone in the field knew, there was no love lost between the one camp (Katzman and Terry) and the other (Glenner). Research skirmishes between the two sides went way back, according to those who worked with them, and proceeded to escalate once they were within close proximity of one another.

In April of 1991, Glenner, who was nearing his sixty-fourth birthday, allowed himself to be persuaded by Joy that he was working too hard and needed a few days off. "Glen had been looking pale and haggard, and I thought it was because of the long, hard hours he was putting in," remembers Joy. They went to stay at an inn in Idyllwild, a small town in the mountains above Palm Springs, and were so taken by the peace and quiet, they decided to scout out the countryside for a vacation home. While driving around with a realtor, Glenner began having trouble breathing and assumed it was because of the high altitude. That night his shortness of breath worsened, and, returning to La Jolla the next morning, the Glenners went straight to the UCSD Medical Center where Glenner, a staff member, was assessed by colleagues. They ruled out a heart attack, but his lungs, it was discovered, contained excessive fluid, a common sign of congestive heart failure. He was tested further, and in a letter later written to one of his favorite former postdocs, Mark Peppys, Glenner described the diagnosis, which he had double-checked for himself by viewing the results of a Congo-red-type staining of his heart tissue. A "cardiac catheterization biopsy," he wrote, "revealed to the stunning shock of everyone—amyloidosis."

Or senile cardiac amyloidosis, as the fatal illness is formally referred to. Colleagues of Glenner's say it was as strange a twist of fate as they have ever known. The primary culprit in this particular amyloid disease is transthyretin, an amyloid-prone protein which, when it aggregates into amyloid fibrils, shares the same beta-pleated, tough-as-nails structure as Alzheimer's A-beta fibril. In Glenner's case, the vessels and chambers of his heart were becoming increasingly clogged with aggregates of transthyretin. While working at the National Institutes of Health, he had isolated this same amyloid protein from patients with the same condition and studied it relentlessly. Upon being diagnosed in the springtime of 1991, he knew full well that he had but four to eight years before the disease ran its course.

An inherited form of this type of amyloid condition exists, but Glenner had no known family history of the illness nor did his DNA reveal any mutations. Possibly his was a sporadic case, Glenner decided, and he'd acquired it during his lifetime—non-infectiously. With the exception of the prion diseases, no amyloid disorder has ever been proven to be infectious, and, according to Joy, Glenner was convinced he hadn't contracted his amyloidosis in the lab. But his family and friends had to wonder.

His illness made him tired enough that he had no choice but to cut back on his working hours. For the next few years, nevertheless, hardly a day passed when he didn't have his nose to the bench or wasn't attending to the needs of patients at the Glenners' day-care facilities. Observes Joy, "He wanted to make every second count."

"Dear Don," Glenner wrote in July '91 to his colleague Donald Price at Johns Hopkins. "Many thanks for your kind note and concern. As you probably know, I have 'senile' (I hate that word) cardiac amyloidosis confirmed immunohistochemically. A little daily diuretic pill keeps the symptoms in check. Needless to say, I am at work daily, but don't whiz around the lab as in former days. There's too much excitement here to stay away."

Whereas Julia had understood so little about her illness, her ten children knew its name, knew its damage, and had a dim sense that they were genetically susceptible to their mother's rare form of the disease. In 1991, their ages, which ranged from twenty-seven to forty-nine, nearly paralleled the years during which the disease's early onset usually struck. They had read that mutations connected to it had been found, but that didn't make their at-risk situation necessarily any more real. No certain signs of it had appeared among them, and they lived their lives around other thoughts—their attachments, their marriages, their careers, and the sons and daughters they were bringing into the world. They'd been told that the disease mostly hit twins; perhaps it would travel no further than their mother and her identical twin Agnes, who had developed Alzheimer's some ten years after Julia. Perhaps the Tatro twins' affliction had been some sort of genetic fluke. Maybe it had been only that.

— *eight* —

Of Mice and People

As he peered ahead into the great land that stretched before him, the way seemed long. But the sky was bright, and he somehow felt he was headed in the right direction.

—E. B. White, *Stuart Little*

By late 1991, it felt as though our field had entered a maelstrom. The discovered mutations on APP were a terrific inducement to chase down genes associated with other cases of Alzheimer's. They also intensified the inquest into the excised A-beta fragment's toxicity and the search for the seemingly delinquent secretases that cut A-beta loose. APP was becoming one of the most pawed-over genes in the entire human genome. The investigation surrounding it and its processing of amyloid would be cited by the publication *Science Watch* as 1991's "hottest" corner of biology.

Meanwhile, tangle researchers had carved their own significant benchmark. Earlier that year, extending work done in other labs, neurobiologist Virginia Lee, pathologist John Trojanowski, and coworkers at the University of Pennsylvania School of Medicine had managed to purify, solubilize, and sequence a short strip of the tangle filament and thereby secure unequivocal proof that an insoluble form of the protein tau indeed was the tangles' base unit. "They nailed it. They gave it the coup de grâce!" observes Dennis Selkoe. More than ever, tangle disciples had a strong gut feeling that they were onto something. Never mind all the prattle about amyloid. Gone awry, tau might be much closer to the events in Alzheimer's that destroyed nerve cells.

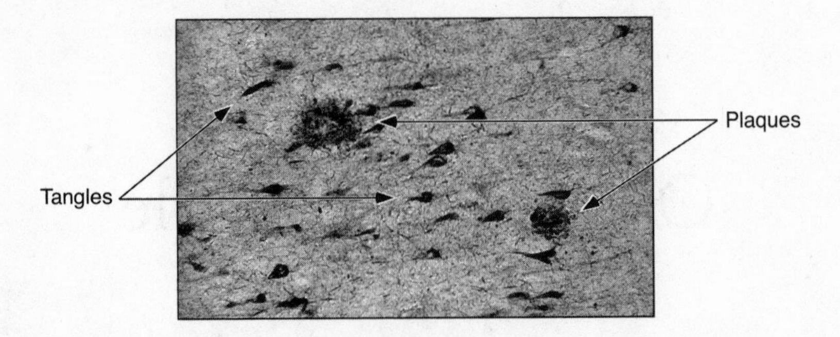

FIGURE 8.1 Photomicrograph of cortex pathology.

As if all these important advances weren't whipping up enough activity, yet another development was brewing. For several years researchers had been striving to fashion the field's first transgenic mice—mice whose genomes were manipulated to create the same plaque-tangle pathology in mouse brains as seen in human brains. ("Transgenics," as the word implies, involves transferring genes from one organism to another.) Left to their natural devices, mice don't develop either plaques or tangles. But it was imagined that genetic engineering might foist Alzheimer lesions upon rodents, by either of two approaches. The first was to insert a normal human APP gene into mice and prod it to overexpress its protein, resulting in too much A-beta and multiple amyloid lesions. The second, which aimed for the same outcome, was to insert a *mutant* human APP gene into mice.

By late 1991, there was much ado over the reports of three seemingly successful Alzheimer mouse models. If transgenics' sleight of hand had worked, these mice might be worth their weight in gold. Their developed lesions could be direct proof that the inserted human APP gene caused the ensuing pathology. One might even glimpse associated signs of dementia. Not that researchers altogether knew to what degree mice got dementia, or if they were capable of getting it, or whether it could be accurately assessed if they did get it.

But the most valuable consequence of transgenic mice was that they might serve as indispensable living test tubes by which to gauge whether a drug compound could rid the brain of plaques and tangles. They might be a conduit to science's end goal—an effective treatment or cure for the disease. This assumed that either the plaques or the tangles, or for all anyone knew both lesions, brought on neuronal death.

For decades, researchers who were curious to learn whether animals besides humans were vulnerable to Alzheimer's lesions had been occasionally using the light microscope to probe animal brains. A handful of accounts had noted Alzheimer-like plaques in older dogs and horses. In the early 1970s, a newer generation of investigators took advantage of the exceptional powers of the electron microscope with the same goal in mind—notably, in this country, Robert Terry and Henry Wisniewski. Before his death in 1999, Wisniewski recalled how, in the early '70s, a conversation he had with the great British pathologist Bernard Tomlinson—one of the trio to have driven home the reality that older and not just younger people got Alzheimer's—persuaded him to study the brain tissue of, in particular, older dogs.

"Tomlinson was a great animal lover. I expect he still is. He assured me that not only did old dogs have plaques, but they also could show signs of dementia. He told me that he'd had a dog who at sixteen definitely had dementia. How did you know? I asked him. He said, 'It came from living with the dog and knowing its habits. It had been a very well-trained dog, very clean, and very well behaved. And slept during the night. Then he stopped sleeping through the night, began bothering us, walked around the house, didn't keep himself clean.' The worst was that Tomlinson's wife was a superb gardener, and they had an impeccable English garden. The dog knew each of the paths, and not to touch any of the flowers. But the dog started to step on all of the flowers, misbehaving totally. He began messing on some of the paths, and then he began getting lost in the garden and couldn't find his way home. Tomlinson told me, 'You'd better look closely at a dog's brain to see what's happening.'"

Terry and Wisniewski confirmed that the cerebral cortex of older dogs as well as monkeys unquestionably could be as plaque-strewn as the human cortex. As for brain cells marred by tangles, the researchers didn't find them in dogs, and only glimpsed "exquisitely rare" instances of tangles in monkeys, according to Terry.

In time, advances in combining electron microscopy with immunocytochemical techniques—the use of antibodies and stains to identify proteins—would part the curtain even further into an Alzheimer-like equivalency in the animal world. For the past two decades veterinary pathologist Linda Cork, the chair of the Department of Comparative Medicine at Stanford University, has applied these tools to brain tissue from some 200 animals in their dotage, and very interestingly in her own words this is what she has found: "Virtually all apes and Old World and New World monkeys can get amyloid neuritic plaques, and their plaques can be as profuse as what one sees in humans," she recounts. "Most old large carnivores develop neuritic plaques, too—tigers, lions, spotted hyenas, leopards, and bears. On the other hand, I haven't seen them in large herbivores like giraffes, hippopotamuses, or donkeys—but, then again, we see relatively few large herbivores at autopsy. Nor have I found plaques in avians, and I've looked at a variety, from aged cockatiels to a twenty-six-year-old flamingo. But, again, these findings come from a small sample size of each species. Nor do plaques appear in smaller animals like mice, rats, African hedgehogs, or mink."

Intriguingly, Cork's observations have led her to believe that for plaques to develop, "a species has to live longer than nine years and have a fairly developed brain that is more like a carnivore's than a herbivore's. That is, the differentiation of the cortex has to be relatively sophisticated," she explains. As for neurofibrillary tangles, Cork had found them only in aged brown bears and polar bears—"and possibly one monkey," notes Cork. "Bears can develop very dramatic plaques and true tangles." Other researchers say they've spied the beginnings of tangles in certain larger animals and theorize that if these animals lived longer, full-blown tangles might occur.

In enumerating this fascinating detective work, Cork emphasizes a major caveat, as she sees it: "You *can* say that many larger animals get brain deposits of amyloid, but you *can't* say they necessarily have Alzheimer's disease."

It's widely accepted nowadays that older, larger animals can slip into the mental numbness of dementia just as people do. When I was a boy, our family had a cat—Beanie, a beautiful gray calico—who after reaching the ripe age of twenty-one, just like Tomlinson's dog, started sleeping fitfully, displaying a marked change in petsonality, and wandering periodically. One day she disappeared for good, leaving a large sad hole in our family. Anecdotal accounts abound that some animals reach the point where they no longer seem to recognize their human friends, and Beanie had seemed close to that state. Whether she had Alzheimer's disease, specifically and in the human sense, is impossible to say, especially since dementia in animals, just as in people, can be caused by numerous disorders. However, had autopsied tissue shown abundant amyloid in the brain, I'd be inclined to think she did have a feline form of Alzheimer's.

Donald Price at Johns Hopkins, a neuropathologist well known for his extensive study of aging primates, believes his research points to a relationship between the advance of amyloid pathology and declining mental behavior. Notes Price, "In older monkeys first we see behavior deficits, then diffuse plaques, then more neuritic plaques, and then, in some, we even see the first stages of tangles." The early behavior deficits, in my own opinion, would seem to go hand in hand with soaring amounts of toxic A-beta fibrils collecting in the brain—before widespread plaque formation occurs.

That three Alzheimer mouse models—not just one—had been independently fashioned by three teams by the end of 1991 raised all the more expectation that the field was gearing up for drug-testing. Two of the mouse models reportedly exhibited amyloid clusters, while a third, amazingly enough, was said to contain Alzheimer's brain pathology in full bloom: both plaques and tangles. These "unfortunate mice," as the *Economist* referred to them, were part of a nationwide trend, for very noticeably medical research was entering the "Decade of the Mouse." Since mice reproduce prodigiously and quickly, and cost less to breed, feed, and house than rats and primates, they were rapidly becoming the animal of choice for myriad biological experiments.

Today, I sometimes feel we've become a bit too trigger-happy when using transgenic mice in research. The growing numbers expended are pretty startling. Based on a survey, the National Association for Biological Research estimated that during 1998 U.S. researchers made use of some 17 million mice, that figure expected to rise by at least 50 percent in three to five years. It used to be that you could get a grant to investigate the role of a gene defect in a disease process just by showing a mutant gene's activity in cells in a dish, but now grants are tough to come by unless you've already commenced mouse studies. Clearly there's a need for transgenic animals for elucidating the biological mechanisms of disease and for developing new drugs, but I think it's a dangerous habit if, by default, we treat mice as though they're merely inconsequential cells in a dish. In instances where we can learn from cells, we should do so, and save mice for experiments that cannot possibly be performed in cells.

The maelstrom engulfing the field, then, wasn't just about the recent advances, it was also about the building confidence, as betokened by work on transgenic mice, that one day perhaps not too terribly far off a treatment truly might be made to delay or even stop the onset of Alzheimer's. (Reversing the disease's course and returning a patient to normal was still viewed as a long shot, since there might not be enough healthy neurons left to fuse a recovery.) Very gradually, the sluggish drug-discovery wheels of large pharmaceuticals were being set into motion by the feasibility of aiming drugs at the recently discovered biotargets. A-beta peptide's possible toxicity, the brain's inflammatory response—even the hypothetical secretases—might serve as bull's-eyes.

The surge in amyloid research had all but supplanted work related to the "cholinergic hypothesis," the idea that a deficiency in the neurotransmitter acetylcholine might be a major contributor to Alzheimer's. The only theory about Alzheimer's causation to have previously galvanized the field, this Lone Ranger of explanations had been slipping in credibility ever since it had been realized that many other brain chemicals were adversely affected by the disease.

Nevertheless, in 1991—a full fifteen years after the advent of the cholinergic hypothesis—numerous designs for cholinergic drugs aimed at preserving acetylcholine were at various stages of development. How long the road from concept to marketplace! Since no drug treatments had been approved for Alzheimer's, many patients, their families, and their doctors awaited these therapies with great hope and anticipation. They just might work, at least in terms of keeping patients in the early stages of the disease from getting worse. Yet compounds based on the cholinergic hypothesis, most drugmakers realized by now, conceptually fell far short of what those based on the "amyloid hypothesis" might accomplish. This term, appearing in the early '90s, sprang from the belief

Alois Alzheimer,
1864-1915

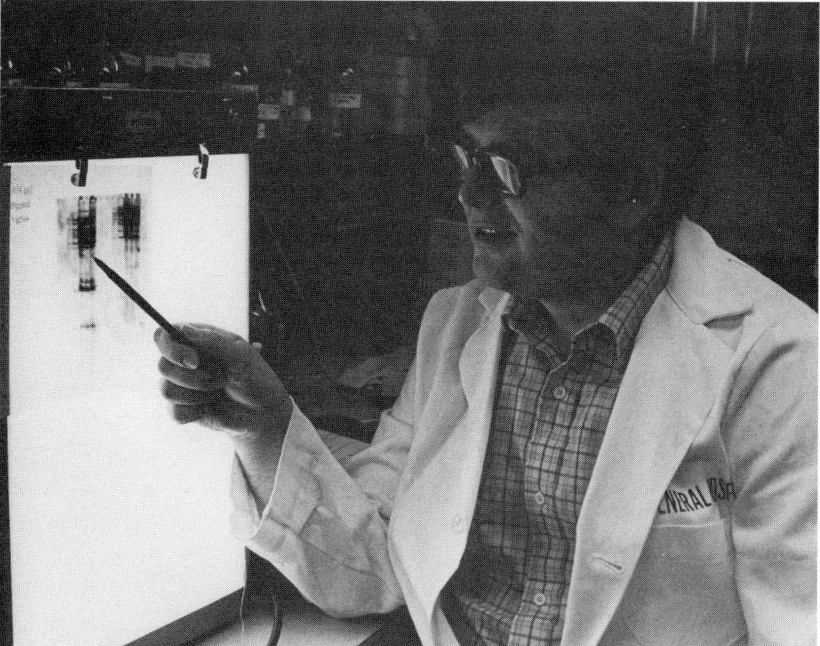

Jim Gusella, circa 1983. Courtesy of Massachusetts General Hospital.

President Reagan and George Glenner, 1982. Courtesy of The White House.

Rudy Tanzi and Henry Wisniewski, 1998.

Stanley Prusiner and neuropathologist Don Price, a leading designer of animal models for neurodegenerative diseases, 1993.

John Hardy, 2000. Photo: Dora Kovacs.

The Alzheimer field's two Bobs: Robert Terry (left) and Robert Katzman. Also neurobiologist Katherine Bick, a fluent chronicler of the early history of Alzheimer research, year 2000. Courtesy of Katherine Bick.

Virginia Lee and John Trojanowski, 1998. Courtesy of Virginia Lee and John Trojanowski.

Bruce Yankner, 1997.

University of Washington/Seattle researchers, 1996. (left to right) Tom Bird, Jerry Schellenberg, and Ellen Wijsman. Courtesy of Metropolitan Life Foundation.

(left to right) Dennis Selkoe, Blas Frangione, Carmela Abraham, and Henry Wisniewski in Canada for an amyloidosis meeting, 1993.
Photo: Carmela Abraham.

(left to right) Ashley Bush, Steven Younkin, Colin Masters, 1991.
Photo: Carmela Abraham.

The Triumvirate: (left to right) Sam Sisodia, Steve Wagner, Rudy Tanzi, 1999.
Courtesy of Dora Kovacs.

The Noonan family, 1999. (Back, left to right) Malcolm, John, Bob, Eryc (front) Maureen, Patty, Julie, Kathi, Fran. Courtesy of the Noonans.

Various members of Mass General's Genetics and Aging Unit, 2000. (left to right) Rob Moir, Tae-Wan Kim, Donna Romano, Rudy Tanzi, Dora Kovacs, and Wilma Wasco. Courtesy of Dora Kovacs.

that neuronal death in Alzheimer's was a direct result of the rampant, accumulating A-beta fibril and its aggregation into amyloid.

Besides being spurred on by biotargets that might interfere with the pathology's very beginnings, drugmakers were feeling the pressure of a significant obligation. A new report suggested that Alzheimer's claimed even more victims than the surprisingly high estimates dating from the 1970s. In 1989, a six-year study of East Boston's population, which had been carried out by a Harvard Medical School team led by Denis Evans, provided evidence that as many as 10 percent of East Boston residents over sixty-five, and *nearly 50 percent of those over eighty*, had probable Alzheimer's. In response, the National Institute on Aging (NIA) immediately had hoisted its estimated numbers of Americans living with the disease from 2.5 million to a staggering 4 million. Presently, the data showing that nearly one-half of the elderly population falls prey to Alzheimer's remains controversial, notes Mass General neuropsychologist Marilyn Albert, one of the paper's authors. "Many other studies have found lower prevalence rates—though none lower than 25 percent," says Albert. "What everyone agrees on is that the prevalence of Alzheimer's goes up quite dramatically with age."

Concurrently, autopsy-based studies of younger people who were heritably doomed to get the disease but died before it struck were revealing the wispy beginnings of brain plaques and tangles ten, even twenty years before the disease's outward symptoms would have arisen. If a drug was to keep the creeping pathology at bay, someone genetically at risk might have to take a daily pill for many years, it was being realized. Hence, from a strictly business standpoint, an effective drug could ring in plentiful profit. It easily could be a "blockbuster," as big-time compounds are called. "Even a kid can do the math and appreciate the big number," says one drugmaker whose company started moving toward the design of Alzheimer's drugs in the early '90s. "Multiply the 4 million people in this country who get Alzheimer's by the annual cost of such a drug—say, to be conservative, $1,200. Then multiply that figure by fifteen years, the length of time a person might need to take the drug." Most calculators don't have room to display all the zeros, the sum total being some $72 billion, or $4.8 billion per year.

In 1981, in an essay published in *Discover* magazine, Lewis Thomas had insightfully declared Alzheimer's to be "not a disease-of-the-month but a disease-of-the-century." "It is the worst of all diseases, not just for what it does to the victim but for its devastating effects on families and friends." Here it was a decade later and the entire general community, confronted by mushrooming reports of lives ruined by Alzheimer's and scientists' attempts to fathom it, was catching up to this perception, equally alarmed. Otto Preminger in 1986; Rita Hayworth in 1987; Sugar Ray Robinson in 1989; Aaron Copland in 1990; Dana Andrews in 1992; frequent notices of well-known figures dying of Alzheimer's and innumerable accounts of ordinary citizens daily seared the reality of the disease more deeply into the national consciousness.

The acute needs of patients and their families, the public's growing awareness, and scientists' expanding clues convinced the federal government to devote all the more resources to battling Alzheimer's. From 1980 to 1987, funding provided by the National Institutes of Health for Alzheimer research had climbed from $13 million to $75.8 million. This hike, granted, was part and parcel of the terrific across-the-board increase in government support of medical research seen ever since the early 1970s. However, from 1988—the year after the finding of the APP-amyloid gene—to 1992, the NIH's allocation would jump even more impressively, from $84.5 to $278.9 million.

In 1986, President Reagan had signed legislation that created a federal advisory panel on Alzheimer's disease, and in 1991, after careful review, the panel recommended that Congress allocate $500 million annually for Alzheimer research. At present, our field receives in the neighborhood of $400 million in federal funds, and we look forward to the day when Reagan's goal is realized. In a perfect world, research allocations should expand exponentially: You get a grant for a project, it allows you to elucidate more clues, which require further investigation and build an argument for receiving even more funds. It being an imperfect world, in most labs dozens of worthwhile projects sit on back burners, awaiting more money and time, in that order.

Compared to funding for wet benchwork, only a minimal number of federal dollars go toward the research, development, and teaching of methods used in caring for Alzheimer patients, according to Paul Raia at the Massachusetts Alzheimer's Association. "Already we can delay a patient's institutionalization by almost a year through certain care-giving techniques, which no treatment can do yet," notes Raia. Raia, the pioneer behind "habilitation therapy" for dementia patients, stresses that there are behavioral-management techniques and therapeutic activities that can as much as "realign the planets" for these patients—mold life around their needs—in order to "maximize their functional independence and morale." In the meantime, that "so little federal funding goes toward care-giving amounts to flagrant neglect," Raia maintains.

The recent discoveries were pumping so much adrenaline into the field that a noticeable effect was the growing division between those of us who espoused the amyloid hypothesis as central to the disease's destruction and those who instead put the blame on corrupted tau and the tangles. At conferences, amyloid-versus-tau debates were flaring up with increasing regularity. Observing the fireworks, an editorial planted a sardonic cartoonlike sketch in people's minds when it labeled these opposing persuasions "Baptists" and "Tauists"—the moniker "Baptist" derived from beta amyloid protein. Inside the field, the names made for some amusement. As Peter Davies wryly expounded at a meeting some years later, "I might as well declare my religious affiliation up front. I'm a Roman Catholic, and not a Baptist and not a Tauist!" Others similarly unattached to either persuasion declared themselves "agnostics."

Yet the Baptist-Tauist portrait struck a certain chord. Amyloid and tau researchers weren't only butting heads at meetings, they were caught up in a tug-of-war for available grant monies as well. Those of us investigating the APP-amyloid gene and its protein, who looked upon Alzheimer's as an amyloid disease, were sailing high because APP's revealed mutations legitimized our work. The mutations argued for more research funds. The tau story having recently made headway, tangle detectives similarly were full of optimism. They too had a long wish list for grant funds. Their diggings had revealed that in the disease too much phosphate attaches to tau, and a cottage industry of inquiry had sprung up over whether this abnormality led to cell death. The tau gene still hadn't yielded any mutations, yet as always Tauists felt on the brink of linking tau more convincingly to the disease.

Raising expectations even higher for both Baptists and Tauists, here were three promising transgenic mouse models—not just one, but *three*.

And what about that mouse that reportedly displayed both plaques and tangles? Some mouse! This particular model had been fashioned by researchers affiliated with the NIA's Gerontology Research Center, Mount Sinai School of Medicine, and Yamanouchi Pharmaceutical. Maybe the mouse's two-pronged pathology was validation that a mutated APP gene, along with resulting plaques, also affected tau in cells and brought on tangles. After all, early-onset patients whose disease was caused by mutations in the APP-amyloid gene got both lesions. What a stellar proving ground this mouse might be!

Or would it? In late '91, when the model's report in *Nature* landed on the desks of senior researchers across the country, you could all but hear a collective cry of "Foul!" Something was terribly wrong about the report. In my own lab, as I studied the text and accompanying photomicrographs, I couldn't believe what I was seeing. The photographed lesions perfectly resembled those in Alzheimer's disease, but it was the neurons themselves that gave one pause. To me, neurons in a rodent brain look like little footballs, yet those in the report looked more like little pyramids, just like the pyramidal neurons found in the cortex of humans.

"When I saw the original," recounted Henry Wisniewski, "I knew it was not mouse brain but human, because of the density of neurons." Recalls John Trojanowski at the University of Pennsylvania, "Virginia [Lee] and I were looking at the pictures together, and we kept saying to each other over and over and over, 'I can't believe this!'"

Allegations quickly mounted that the NIA scientist responsible for gathering the rodent tissue had substituted lesion-replete autopsied brain tissue from Alzheimer patients. The researcher was new at the NIA, and many believe that the pressures of performing simply may have gotten to him. "Like everyone else, scientists can develop psychotic episodes," noted Henry Wisniewski, who was one of the first to bring the matter to the attention of the NIH. "In science, if you do something wrong, you're immediately found out. It's very different from politics, where humanity pays the price of Hitlers. When politicians acquire the power, no one has the strength to turn them down until it's too late."

Suffice it to say, a scientist's fall from grace is never pretty. "I was dumbfounded that he threw his career away," mentions one researcher. The rumblings after this incident included the objection, How did the report, which soon was retracted by its authors, get into print? Why didn't its peer reviewers notice something amiss? But this is easier said than done. When scientists serve as reviewers and evaluate colleagues' work, we usually take the data at face value. If the data warrant, we can be incredibly critical and questioning, all the more so when a paper's contents are controversial or highly significant. But we don't expect wrongdoing, and our eye doesn't necessarily catch the type of detail that would give it away.

By mid-March of 1992, yet a second mouse model had been retracted. Developed by scientists at Miles Research Center, it too had a serious problem, albeit a legitimate one. The rodents' brain deposits were growths that naturally arose in the mouse strain used by the researchers, it turned out. With two mouse models scrapped, that left a third developed by Barbara Cordell and coworkers at California Biotechnology (now Scios, Inc.). Many felt this attempt fell short, however. The mouse's limited number of brain plaques was equivalent to what one saw in a normally aging human brain, and therefore a weak model for testing a drug's ability to stifle plaque buildup.

Backed by drug companies whose interest in Alzheimer's was on the upswing, the push toward transgenic animals nonetheless would continue. The goal of getting mice to develop the disease's blackguard lesions for the testing of potential inhibitors was far too important not to keep trying.

In the genetics arena, after the sudden progress attached to the APP gene, there'd been an equally sudden lull. APP mutations accounted for so few cases of Alzheimer's, an enigma shone all the more brightly. If not APP, what other genes lent to Alzheimer's preponderance of cases? And how did these genes press APP's button for the release of its seemingly egregious, aggregating A-beta peptide?

Malcolm—Julia's sixth-born—remembers too clearly the June day in 1993 that he and his wife visited an older sister and her family at their lakeside cottage in New Hampshire. Little about Fran—Malcolm's forty-four-year-old sister—resembled their mother. She had their father's pale blue eyes and an aptitude for gadgets that no one else in the family possessed. She was considered the family technician, since she could do everything from rewire a house to repair an IBM hard drive. With an IQ over 150, she was sharper than a tack, which is why Malcolm felt a fear rip through him when, for the second time that day, Fran picked up the blue-green embroidered bracelet off the hall table and wondered aloud where it had come from, and her daughter had to tell her yet again that she'd made it for Fran. Later, Fran would ask a third time. Whose bracelet was this? Since their mother's death, it was as though the disease had been secluded in a crevice between generations. But now it had crawled back into their family's midst, Malcolm realized, never having left.

— *nine* —

Gene Prix

What price the miles thrown away if we can start the sheets and burn out a bearing in the patent log?

—Alfred Loomis, "Ocean Racing"

Spring was on the move in Seattle. Icy rivulets streamed down the sides of Mount Rainier, Mount Stuart, Mount Snowking, and their snow-capped brethren, feeding into rivers like the Nisqually, Cedar, and Skykomish, which swelled and rushed headlong into Puget Sound. In thousands of warming ponds and lakes, tadpoles wiggled out of the bottom ooze, their set of tens of thousands of genes in each cell encoding a representative throng of proteins, which, moment by moment, changed the tadpole a bit more into a frog. Princes, politicians, cooks, golfers, and scientists are made in similar fashion.

On a late June day in 1992, Jerry Schellenberg stood idly beside a printer in the University of Washington's School of Medicine waiting for the numbers to emerge from a recent genetic analysis of nine of his lab's early-onset Alzheimer families. Ever since 1985, Schellenberg's team had been scanning its growing collection of Alzheimer DNAs for a mutation, but the results had been unrevealing and the chore of analysis had grown more and more tedious. Sixty-three markers from the genome had been tested, with special interest paid to chromosome 21 and more recently chromosome 14. No markers having panned over the years, Schellenberg didn't expect that the sixty-fourth would either. A consummate neurogeneticist and refreshingly sincere, Schellenberg was someone whose friendship I'd come to greatly value in a field that seemed increasingly prone to secrecy and the backlash of secrecy—paranoia. He and I regularly chewed things over—findings of whatever size, as well as dead ends.

"So much time had gone by without getting any positive returns that I'd turned blasé," recalls Schellenberg. "As the printer printed out the numbers, I was staring at the wall in a fog. My eye fell and suddenly positive numbers intruded. No negatives! We had a hit!" *Chromosome 14 had something on it.* "All I kept thinking as I bicycled home that night was that I hoped I didn't get into an accident and die before I could tell my wife. It was such a weird thought, because I never think about dying."

For years, chromosome 14 had been behaving like an ambivalent seductress. It would flirtatiously imply it harbored an Alzheimer mutation, only to dash hopes. Faint signs of an anomaly far down on the chromosome had been seen as far back as 1983 but never led to anything. The chromosome once more had fetched interest in the late '80s. Like Rachael Neve and me, biochemist Carmela Abraham—a Harvard classmate of mine working out of Hunt Potter's Harvard Med School lab—had gone after the amyloid gene via its protein. In the process, much to her own surprise and everybody else's, instead of pulling the amyloid protein from plaques, she'd sequestered a protein named ACT—for alpha1-antichymotrypsin—whose gene resides on chromosome 14's long arm. It wasn't nearly so abundant in the plaques as amyloid, but its presence in Alzheimer brains was riveting for several reasons. First, the ACT gene resided in the old suspect region on 14. Second, its protein was a protease inhibitor. Therefore, maybe it hampered beneficial proteases that break down A-beta, and thus spelled trouble. Third, because ACT was known to be involved in the body's inflammatory response, its occurrence in plaques supported a role for inflammation in Alzheimer's, notes Abraham, who now runs her own lab at Boston University's School of Medicine. ACT in plaques, therefore, was of the utmost interest to those researchers exploring inflammation's contribution to Alzheimer's amyloid.

Yet chromosome 14 again spurned hopes. Its ACT gene yielded no clear-cut evidence of an early-onset Alzheimer defect. The chromosome's south section was so poorly mapped back then, it discouraged further rooting about, and once more activity on the chromosome largely fell silent—until 1990, which is when different sorts of variations in DNA began coming into view. A godsend, they promised to be far more informative than the single-base type we'd been using to track mutations. Unmasked by James Weber's group at the Marshfield Research Foundation in Wisconsin, they amounted to short repetitive DNA sequences and hence were named "short tandem repeats" or STRs. Led forward by these "genetic stutters," gene hunters could analyze formerly impassable stretches of chromosomes and better compare chromosomes in each inherited pair to root out discrepancies.

Those first found in the genome and used as markers were dinucleotide repeats—repeats of two bases, such as CACACACA. In a few years we'd be employing trinucleotide repeats, like GTCGTC, then four-base or tetranucleotide repeats. Such stutters occur within both the coding regions (genes) and noncoding regions (junk DNA) of the genome. Why, no one is quite sure. Like single-base variations, perhaps they're leftovers from the remodeling and evolving of DNA that has transpired since even before our distant kin were sea creatures scrounging for survival in the primeval oceans. Whatever their origins, as genomic markers they coincided nicely with the deepening conviction that we in the field should get off our duffs and broaden our search away from chromosome 21.

The sixty-fourth marker—the linkage analysis of which had jolted Jerry Schellenberg out of his fog on that June day in 1992—had been a dinucleotide repeat from chromosome 14, and its coinheritance with the disease in one early-onset Alzheimer family imparted that a mutation lay somewhere within a 30-million-base region in the chromosome's southern section. An even greater thrill for Schellenberg's lab was that analyses of eight other unrelated Alzheimer families lit up the very same chromosome. Their linkage was less robust yet detectable. "We were sure that this locus was going to account for most of early-onset familial Alzheimer's," relates Schellenberg, given that so many unrelated families evidenced it. "One of our big worries had been that nine different mutations might be responsible for the disease in the nine families, so it was a pleasant surprise to think that just one mutation might be accountable for all nine."

But one group of early-onset families tested by the Seattle team—the Volga Germans, which had been gathered by neuroscientist Thomas Bird (as had been most of the lab's Alzheimer kinships)—bore no signs of a chromosome 14 mistake. Schellenberg, Bird, and Ellen Wijsman, who directed the statistical side of the research, couldn't help but be aware that the Volga Germans' mutant gene remained on the loose.

Hurrying to finalize the chromosome 14 analysis and see it into print, Schellenberg recalls the nervous strain of constantly looking over his shoulder and wondering if another team hadn't also hooked onto 14. He was particularly afraid of being scooped at the Third International Conference on Alzheimer's Disease and Related Disorders, its July '92 meeting scheduled to take place in the Italian hot-springs town of Abano Terme. As much as he would have liked staying put to continue work on his team's linkage report, he flew to Venice, took a train to Padova, then a bus to Abano Terme. "I'd slid the data in my shirt pocket and was all ready to announce our finding in case someone else did, but as soon as I talked to everyone, and it was clear that no one else was going to, I left the meeting"—and boarded an earlier-than-planned flight back to the States. No scientist likes to be preempted!

His group's paper proclaiming an error on chromosome 14 appeared in the October 23, 1992, issue of *Science*, and for a brief, well-deserved moment, the Seattle lab was the sole king of chromosome 14. The media was beginning to back-page any news of a disease's linkage to a chromosome, preferring to save fuller coverage for when the actual gene and its error were flushed out. Nevertheless, the discovery on 14 was trumpeted far and wide. "Science Zeros in on Suspected Gene," beamed the *L.A. Times*. "Alzheimer's Disease; Genetic Link Is Found in Younger People," ran the *Seattle Times*. The next moment, however, official word seeped out: The University of Washington group was not alone. Three other Alzheimer groups were on to chromosome 14, their separate reports slated for the December issue of *Nature Genetics*. The team leaders included John Hardy and his lab partner Mike Mullan, who together had relocated from England to the University of South Florida in Tampa; Christine Van Broeckhoven at the University of Antwerp; and Peter Hyslop, who by then had left Mass General and taken a position at the University of Toronto.

The first I learned of all the hoopla surrounding chromosome 14 was in late October, shortly after returning from a week's scuba-diving stint in the Sea of Cortez. In the middle of a weekly neuroscience seminar in Building 149, Jim Gusella leaned over and whispered in his nonchalant way that Jerry Schellenberg's and Peter Hyslop's labs had evidence of a major early-onset Alzheimer mutant gene on chromosome 14. We were sitting up front, so I couldn't make any wild gestures, but I distinctly remember nearly spilling hot coffee on both our laps.

It was an absolute stunner, particularly the fact that Hyslop in Toronto had linked all four of Mass General's original families to 14, thus confirming that his previous linkage to chromosome 21 was false. Although Hyslop had moved to Toronto in early '91, he'd brought samples of the four families' DNAs with him, therefore retaining a working relationship with Gusella's lab. Here I'd spent years working on the problem of where the genetic error or errors were located in these families, and suddenly it was halfway solved. The defective gene still had to be found, but its true chromosome had been cornered. Like others, I had thought it might take forever to gain clear proof of another Alzheimer-associated chromosome, especially after all the dragged-out testing and ghost-chasing that had occurred on chromosome 21. It was hard to believe that in the short week I'd been in Mexico, obliviously swimming around underwater with parrot fish, so much important news had bubbled to the surface.

From a research perspective, the three chromosome 14 reports that followed Schellenberg's were of the utmost value. They left no doubt that the still-hidden flawed gene on 14 was a major source of the disease's early, vicious strike in many families. Hyslop's account that six unrelated families—the four MGH families and two others—were linked to chromosome 14 was particularly noteworthy in this respect.

Yet once the three reports came to light, a volley of veiled accusations erupted that shook the field far harder and encompassed far more major players than any previous eruption. How was it, it was asked, that so many teams had gotten so lucky all at once? Locating the home chromosome of a disease defect involved a massive time commitment and more than an ounce of luck. Four teams, it was true, had lassoed the APP gene, but we'd made a beeline to the gene from George Glenner's protein sequence, whereas with genetic linkage you didn't altogether know your destination until you got there.

The neck-and-neck timing of the papers created worry and suspicion that a certain amount of piggybacking of information must have occurred. The protests, for the most part, were subtle and hushed. Most individuals didn't confront each other face-to-face, but voiced their complaints to third parties, and the discontent further tunneled through telephone and E-mail, eventually surfacing and blanketing the research community like a wet, solemn fog. The secrecy and behind-the-back tactics developing in the field seemed to be damping lab-to-lab and even in-lab communications that used to be more gracious, open, aboveboard.

The most rumor-driven collision was between Hyslop and the Hardy-Mullan team. Initially, when Hyslop had submitted his chromosome 14 paper to *Nature*, Hardy and Mullan, among others, had peer-reviewed it and, while recommending it, faulted it on the grounds that since Hyslop had previously linked the four Mass General families to chromosome 21, how could he reconcile now linking the same families to a different chromosome? Although several mutant genes might cause early-Alzheimer's throughout the population, most likely one mutation bestowed the disease in a single family. So if the four families indeed had the same defect in common, all four were either linked to one chromosome or to another, but not both.

Nature rejected Hyslop's paper, and Hyslop, rumor had it, felt that Hardy and Mullan had delayed his paper's publication for the sake of completing their own chromosome 14 paper. Hardy tried to clear up the matter. "Rumor is a terrible thing," he wrote in a letter to Hyslop. "We have heard rumors that you have heard rumors suggesting that we unfairly reviewed your ch14 paper to *Nature*." Hardy maintains that, prior to receiving Hyslop's paper for review or even hearing about Schellenberg's paper, his group already was pursuing chromosome 14 because of Carmela Abraham's evidence that the ACT gene on 14 might be Alzheimer-related. His letter to Hyslop continued: "I want us to get on well. . . . At the moment, I am aware that, yet again, AD genetics is besmirched by problems."

Reportedly, even before Schellenberg had published his chromosome 14 linkage, word had leaked out about it. Hardy admits this information made his lab press forward more rapidly on chromosome 14. Some believe the same hearsay may have similarly sped up Hyslop's efforts, who, before he left Mass General, also was working on chromosome 14 because of the suspicious ACT gene. "It wasn't legitimate how people caught on to it," contends Schellenberg. "It was disappointing when these other groups came out, because we had done some superb science, and it should have been out there by itself—although in the long run we did get recognition." Hardy weighs the situation differently. "When you hear a rumor that a mutation is supposedly on a certain chromosome, what do you do? You're spending taxpayers' money!"

Speculation spun over who might have actually landed the chromosome first. Christine Van Broeckhoven believes her Antwerp team did, but she recounts that she had decided to sit on the evidence and find the actual faulty gene first, only to hurry her chromosome 14 paper into publication once she caught wind of the others.

The days of fiercely wishing that inherited Alzheimer's might be easily explained by one flawed gene were definitely over. A lineup was forming. APP on chromosome 21. The at-large gene on chromosome 14. The unidentified Volga German gene. Then that same fall—the logjam autumn of '92—there was news of yet another culprit.

At the Society for Neuroscience's late-October meeting in Anaheim, Allen Roses announced that he and his Duke University coworkers had firm documentation of a late-onset gene on chromosome 19 they'd been pursuing for some years. Greater detail about the gene—apolipoprotein E, or APOE—would appear in several subsequent reports. Since it was the first gene tied to Alzheimer's all-too-common late variety, APOE was nothing short of a blockbuster discovery. The *Journal of NIH Research* referred to it, along with the fresh evidence of a gene on chromosome 14, as the "October surprise"—a surprise in that while so many of us were investigating APP and how it heaped so much amyloid upon the brain, up popped these other disease genes, making it plain that the genetic causes of Alzheimer's still largely eluded the field. The October "riddle" might have been an even better way of putting it. To those of us circling the amyloid hypothesis, APP-derived amyloid seemed the disease's pivotal problem—the disease's desperado. Yet where did the newly sighted genes fit into the amyloid puzzle? Maybe they themselves, through further analysis, would tell us.

On its part, the late-onset gene APOE told a very different story than the apprehended early-onset APP gene. Unlike APP, APOE's harmful hold didn't appear to involve mutations that caused Alzheimer's with all but absolute certainty. Nothing so definitive as that. Like APP and many genes, APOE had alternate forms, each available for inheritance, which differed from one another by just a few bases. APOE had three variants—E-2, E-3, and E-4—and the Duke researchers had found that E-4 appeared to harbor a *polymorphism* that could mean trouble and increase a person's risk for Alzheimer's. (Like a mutation, a polymorphism is an alteration of bases away from the norm. But unlike a mutation, which is both a rarity in the population's gene pool and almost always causative, a polymorphism is common in the population and usually doesn't cause a disease with total certainty.) People dealt two copies of E-4 often, but not always, faced a substantially greater risk for Alzheimer's than those who inherited any pairing of the E-2 and E-3 variants. Roughly 2 percent of the population, it was known, was born with two E-4s. One copy of E-4 also could increase the chance of developing Alzheimer's, but to a much lesser extent.

It's currently believed that APOE is involved in the neighborhood of as many as 50 percent of late cases of Alzheimer's. This doesn't mean, however, that 50 percent of late-onset cases are solely caused by it, since other genetic, aging, and environmental factors likely contribute to these cases.

When this alarming news about APOE descended, it was already known that the apoE protein transported cholesterol and other lipids through the bloodstream, and that in this role its E-4 form also was up to no good. The Framingham Heart Study had correlated E-4 with a near doubling of coronary heart disease in middle-aged women and a 50 percent higher risk in men as opposed to people born with the safer APOE variants. Should a person inherit two copies of E-4, therefore, it possibly incapacitates both brain and heart later in life.

Due to its E-4 variant, APOE represents a *susceptibility* gene—a gene in which a polymorphism increases a person's risk for a disease but doesn't guarantee they'll get the disease. Since APOE-4 is *neither necessary nor sufficient* to cause the disease, like other susceptibility genes associated with other diseases, it's as slippery as an eel and hard to pin down. If it acts up, it seems likely that other genetic and/or environmental risk factors must also be involved, conspiring to bring on disease.

To borrow a model from cancer genetics: Two workers in a factory are exposed to a noxious solvent, whereupon one gets lung cancer and the other doesn't. Odds are, the worker who gets cancer can't tolerate the toxin because of one or two preexisting conditions. He or she might have a predisposing genetic risk factor in his or her DNA for lung cancer, and/or has been exposed to another environmental insult, perhaps at home, that compounds the toxin's dangerous effect on lung cells.

As to be expected, the Duke team's published revelations about APOE got tremendous attention. "Scientists Detect a Genetic Key to Alzheimer's," boomed the *New York Times* on its front page. Among scientists, the example of APOE drove home how many other susceptibility genes might be involved in the disease's late onset. Most genetics teams, including my own, nevertheless stayed focused on the disease's early-onset face. As rare as this face is—presumably as rare as its mutations are in the population's gene pool—each early-onset mutation invariably confers Alzheimer's, which gave us a complete pathological path to try to sort out.

The thinking was this: Divulge exactly how the early-onset mechanism worked, and it probably would be applicable to the disease's older cases as well. Therefore, since the gene hidden in the shadows on chromosome 14 might help explain that black box, it was very much on the hot seat. At the 1992 Neuroscience meeting there was lively speculation about which gene on 14 was *the* gene, as it was being hailed. "Personally, I'm betting on TGF," someone would wager. TGF-beta, a growth-factor gene, possibly contributed to how much APP protein got made, and therefore was a likely suspect. "My vote is with FOS," yet another colleague would throw in—FOS, a global mediator of expressed genes, including APP, drew heavy bets. The truth be told, Alzheimer's biological parameters were so ill defined, you could build a case for just about any flawed gene-protein, usually in less than five seconds.

It wasn't only the new Alzheimer genes that had people abuzz at the '92 Neuroscience meeting—a buzz that carried over to Space Mountain at Disneyland, a favorite pastime for scientists visiting Anaheim. Ever since Glenner had isolated the little A-beta peptide from the plaques, A-beta was imagined to be an errant by-product of APP, a product of the disease. Since the disease's brain plaques were an aberration, their core fibrils must be too, it was assumed. But earlier that fall experiments conducted by four American groups showed that A-beta was, astonishingly enough, a normal output of cellular activity. The teams that independently made this finding, aided in part by improved antibodies, were led by Dennis Selkoe and Christian Haass at Boston's Brigham and Women's, Bruce Yankner at Boston's Children's Hospital, Dale Schenk at Athena Neurosciences, and Steven Younkin at Case Western Reserve University.

Every day of a person's life, they detected, A-beta is constantly made and secreted by healthy brain cells as well as healthy lung, blood, liver, and other type cells all over the body. The field let out a collective *ah-so!* Now we could more accurately gauge the veritable road to amyloid. It wasn't that *dead* cells released the APP protein, which, plaque-bound, got cut down to A-beta. It really seemed as if A-beta must get snipped free from *healthy* cells by the secretases, then free-float, then aggregate into plaques.

Drugmaking was suddenly seen in a whole new light. It was all the clearer that if the production of A-beta was to be prevented, a drug had to directly target the cells as opposed to the plaques. And since healthy cells body-wide were a constant source of A-beta, how convenient for drug testers! "It meant you could start adding drugs to cells and screening them, to see if you couldn't find drugs that reduce the amount of A-beta the cell releases," observes Steve Younkin, currently at the Mayo Clinic in Jacksonville, Florida. Before, testing drugs against A-beta had seemed so improbable, so problematic. "Before, we thought A-beta was only produced if a person's brain was rotting," notes Younkin.

For those of us who saw Alzheimer's as a cerebral amyloidosis, something else crystallized. "Suddenly it became clear that A-beta in Alzheimer's might be acting like cholesterol in arteriosclerosis. They are both entirely normal products, and they're both capable of building up to excess," notes Dennis Selkoe. Although A-beta's release from its parent protein was normal, in high concentrations it might be toxic to brain cells. Since several of the early-onset mutations in APP occurred near the sites where the secretases clipped out A-beta, conceivably they influenced the secretases to liberate abnormally elevated amounts of A-beta!

A drug vision loomed: Similar to treatments for high cholesterol, drugs might be made to reduce A-beta's production, perhaps by blocking the activity of the secretases.

Our altered view of A-beta awoke countless questions. If this fragment was released throughout the body, why did it collect as amyloid only in the brain? If it aggregated in other parts of the body, did certain forms of macrophages—scavenger-type cells not seen in the brain—get rid of it? My growing concern was that if A-beta was normally produced, it must have a purpose, and if you tried stopping its production with a drug, you might run into terrific problems. But that was a risk that the field just might have to take. Moreover, a drug need not turn down *all* the A-beta the body produces.

If only it could be identified, the gene on chromosome 14 might make more sense of the whole picture. The brouhaha over who had piggybacked off whom to get onto chromosome 14 left behind injured feelings but quickly faded as teams shifted into high gear to hunt down the gene. As concerned the DNAs of the thirty or so early-onset Alzheimer kinships stored in Jim Gusella's unit, I felt a special affinity with these families; I'd worked endlessly over them, beginning with the very first one acquired—the Canadians. Tests showed that just like the original four, as many as two-thirds of our other families were linked to the chromosome 14 defect. Preparing to go after the gene, by late 1992 I was shifting grant funds away from chromosome 21 and toward chromosome 14, as well as hiring on more technicians.

In hindsight, the initial linkage to chromosome 14 was simply the starting gunshot. Out of the gates tore various labs, forming partnerships to strengthen their chances. Jerry Schellenberg formed a collaboration with Sherman Weissman at Yale. John Hardy at the University of South Florida and Alison Goate at Washington University in St. Louis, Missouri, joined forces with Christine Van Broeckhoven in Belgium. And, in a semiformal arrangement, Peter Hyslop and I teamed up—a coupling born from the tradition of working off the same family DNAs. When Hyslop had left Mass General for the University of Toronto, Jim

Gusella had approved his removal of half our Alzheimer DNA stocks. As it was, Hyslop took so much of the DNA that it had taken the better part of a year for technicians to regrow the cells and extract enough DNA so that experiments could continue.

An untold number of other groups entered the race, including, rumor had it, Mercator Genetics, Rhone-Poulenc, and various other pharma and biotechs.

The stakes became even more appreciable in the winter of 1992–93 when two groups—Selkoe's Boston lab paired with Ivan Lieberburg's at Athena as well as Steve Younkin's team at Case Western—clinched the first proof that, *yes*, a mutation in the APP gene (the Swedish mutation) resulted in the APP protein's release of A-beta amounts that were many times greater than what APP normally produced. Their tests as good as caught the mutation red-handed. Although confined to the lowly petri dish, it was first proof of a biochemical cause and effect between a mutation in patients and the ensuing pathology. Would we find, as some among us expected to, that a mutation or mutations on the chromosome 14 gene somehow led to the same scenario—APP's turnout of too much A-beta in the brain?

In early 1993, while I was clearing the decks to go after the chromosome 14 flaw, jubilation and celebration all but rocked Building 149's foundation. After a blistering ten-year search, the Huntington's disease gene at last had been snared. The international Huntington's consortium led by Jim Gusella and one of his principal scientists—molecular geneticist Marcy MacDonald—had narrowly inched out two other groups. Gusella's lab having begun the chase, it was only fitting that his unit had presided over the final surrender. In the end, the long-elusive mutation on chromosome 4 sat where it had probably sat since the pharaohs—on a gene near the tip of the short arm. The abnormality boiled down to a strip of DNA that repeats too many times, an extra long "genetic stutter" akin to the shorter ones we had begun employing as markers to find disease mutations.

"Ottawa-Born Scientist Finds Huntington's Gene," blazoned Gusella's hometown *Ottawa Citizen*. *Huntington's Gene Is Found at Last*, trumpeted hundreds of newspapers from Boston to Singapore. For Huntington researchers, ahead lay the even more formidable challenge of using the information obtained from this broken gene and its protein to combat the disease. Lucky for them, Huntington's was a one-gene disease. Alzheimer researchers already had two in the bag, with a third gene and in all likelihood numerous others to be wrestled down.

The Huntington's victory struck many of us as a lucky omen for the tracking of the gene on chromosome 14. The DNA markers guiding us were so improved and so many more Alzheimer family DNAs had been collected for reference that it was reasonable to think that one team or another might pin down the irregularity quite quickly. Granted, in my lab, 30 million bases was a lot of ground for two principal investigators and four technicians to cover. Say that a strand of DNA, instead of being trillionths of an inch thick, is half an inch thick. That meant searching the length of a half-inch-wide ribbon that extends 600 miles, or approximately the distance from Boston to Toronto, to find a defect about as long as a paper clip, assuming the defect amounted to a single base of DNA.

Just as other groups were doing, my lab began by narrowing down the suspect region by way of *positional cloning*. It's a bit like whittling a stick down at both ends. Beginning with pairs of DNA markers taken from high and low on the chromosome's suspect region, we ran them against our family DNAs, looking for any that might segregate consistently with those who had the disease. Ploddingly, we moved from one set of markers to the next, following those that coinherited best with the disease and heading deeper into the region where the gene presumably lay.

By mid-1993, we had narrowed the suspect territory down to about 20 million bases; by the end of 1993, 10 million bases. For the technicians, who were finding markers in DNA and doing most of the sequencing, the work was as monotonous as picking cotton. For me and Wilma Wasco—a biochemistry fellow from MIT whom I'd hired on in the spring of '91—the job of analyzing by eye various DNA markers in dozens of individuals in order to determine which markers were closest to the gene was interminable. We often felt we were chasing something in the dark. Which we basically were.

Before long we would isolate, sequence, and inspect specific genes on the chromosome for mutations. The chore of manual sequencing was so fearsomely slow that as I drove to work on Wednesday, June 23, 1993, stalled as usual in traffic on the Southeast Expressway, I was trying to calculate how the lab could afford a DNA sequencer. This $150,000 machine could save us months of hours of decoding time. I'd switched on National Public Radio and certain of the newscaster's words began sinking in—"preeminent scientist"—"the blast shattered"—"Down syndrome"—"Alzheimer's"—"Epstein, age 59."

The media, that summer, was just beginning to append the name "Unabomber" to an unidentified individual suspected of sending package-concealed bombs to airlines, universities, and computer companies, as well as distinguished scientists, businessmen, and others since the late '70s. (By the time Theodore Kaczynski was taken into custody in early 1996, three would be dead, twenty-three others injured.) My esteemed colleague Charles Epstein, I realized, was the latest target—seriously injured, but thankfully still alive. The head of the Division of Medical Genetics at the University of California/San Francisco and the editor of the *American Journal of Human Genetics*, Epstein was nationally known for his work in both Down syndrome and Alzheimer's. We'd previously published together, including a 1991 paper that described the methodology behind correlating duplicated parts of chromosome 21 in Down patients to the disorder's physical features—the very research that had originally pointed me in the direction of Alzheimer genetics nearly a decade ago.

The explosion that almost took Epstein's life had occurred the previous afternoon. Returning to his home in Marin County, Epstein was in the kitchen opening his mail, which included an eight-by-eleven-inch padded brown envelope. "It was a pipe bomb, and the major blast from it went past my right arm. Had I been holding the package at a slightly different angle, it would have killed me," recounts Epstein. Parts of his fingers on one hand were blown off, and he suffered serious facial and abdominal wounds, a fractured arm, and partial loss of hearing. The kitchen was demolished. In his "manifesto," Kaczynski foresees genetic engineering as eventually turning human beings—"a creation of nature, or of chance, or of God"—into "a manufactured product." A very legitimate fear for sure, but one to kill over?

Epstein's narrow escape put everyone in Building 149 and similar medical-research establishments across the country on immediate alert.

Mail coming into Building 149 was closely monitored for any suspicious-looking parcels. Those of us involved in genetics started thinking twice about having our names in the papers, because of the visibility. Like many in the building, Donna Crowe—the office manager for neurologist Bob Brown's neuromuscular lab—remembers it as a scary time. "I used to joke with Bob that since I opened the mail, I was the sitting duck, not him." But, since it was no joke, she opened the mail each day with a good deal of caution.

Arriving at work one morning, I ran into Jim Gusella in the parking garage, who, for a big guy, was carrying a padded envelope out in front of him ever so delicately. "It was sent to me at home, and I'm not sure I trust it. I'm handing it over to security"—and off he went, plainly preoccupied. When I got to my office and was going through some mail I'd brought from home, I discovered the exact same item. Same shape, same thickness, same type of address label. After consulting Wilma Wasco, I decided to rip it open as fast as possible, which, as Wasco quickly backed into the hallway, is what I did. The contents turned out to be a videotape from the store Tweeter Etc. advertising a surround-sound home-theater system. Gusella's package was the very same.

Our digging on chromosome 14 continued. We'd begun isolating known and unknown genes in the region and testing them for defects by comparing bits of their sequences in our diseased and non-diseased family DNAs. But despite the optimism that the chromosome's flaw would be a fast find, 1993 was turning out to be about as uninspiring as a 1991 Bordeaux.

When the yearly Society for Neuroscience meeting rolled around in the fall of 1993, there was terrific anticipation among our research crowd about an NIA-sponsored seminar in which Allen Roses and his Duke colleague Warren Strittmatter were slated to share their theory about how APOE—the risk-factor gene unearthed by Roses's lab—might cause late Alzheimer's. The room was jammed, since for months the phrase going around was that the Duke scientists were going to "lay a bomb" on our research community.

But to my ears at least, no detonation ensued. Their hypothesis wrapped around the idea that APOE-4—the gene variant they'd shown raised the risk of late-onset Alzheimer's—promoted tangles in an indirect way, and thus neuronal death and dementia. Their test-tube studies suggested that the apoE protein's safer versions—E-2 and E-3—bound more tightly to the tau protein than E-4, suggesting that E-4 had less of a stabilizing effect on a neuron's infrastructure. What seemed improbable to several of us about this tangle-heavy theory was that the experiments had involved much higher concentrations of proteins than one sees in the body. We Baptists had an even more general complaint. The tau protein is a component of a cell's skeletal structure, while apoE gets secreted outside the cell. Existing in such different places, how would the two possibly interact?

Some, however, applauded the Duke team's theory—Zaven Khachaturian, for one, the associate director of the NIA's neuroscience aging program and a guardian angel who had helped procure many a grant for Alzheimer researchers. "By putting forth such a provocative idea, Allen shook the field, and I think that's healthy," Khachaturian stated in *Science*. Like others, Khachaturian felt too much emphasis was being placed on amyloid research, to the detriment of other roads of inquiry. "Such scientific orthodoxy is pretty dangerous," he warned.

My own skepticism about some of the claims regarding APOE and its role in Alzheimer's was putting me at odds with Allen Roses. In particular, I had to wonder if APOE-4's polymorphism was a bad actor all on its own. For it seemed possible instead that it might travel with, hence be a marker for, a nearby discrepancy, one that lay elsewhere in APOE or even in a nearby gene.

The greater the progress, it seemed, the more it led to strained relationships and opposing viewpoints. In 1993, teammates John Hardy and Mike Mullan had an especially rough time. They were members of the St. Mary's lab that had tracked the APP gene's "London mutation"—the first mutation associated with Alzheimer's. Hardy, Mullan, and others in their group had left behind England's struggling science community and made good on an offer to bring their lab to the University of South Florida's Department of Psychiatry. Ever since World War II, funds for scientific research in England had been shriveling, and the St. Mary's researchers had found themselves and their families squeezed between meager salaries and soaring mortgage rates.

Once in Florida, Mullan expressed his relief to a London reporter: "Now I can be a researcher not a beggar." He and Hardy were considered two of England's finest scientists, and their leave-taking would push Parliament to try to determine how to stop England's "brain drain."

The two had worked well together in England, but once in Tampa "there was enormous pressure on both of us to have a hero—for someone to have the power and control in our group," recalls Mullan. This is less true in England, where lab responsibilities are more equally distributed. Hardy's sometime carelessness in England, says Mullan, contributed to Mullan's handling of the "Swedish mutation"—yet another mutation on APP that Mullan and labmate Fiona Crawford had isolated after they'd arrived at the University of South Florida. They had it privately patented—"because I'd found it, and because John hadn't been convinced the Swedish family even had a mutation," notes Mullan. However, this didn't sit well with Hardy, who'd originally arranged for the Swedish family's DNA analysis. Patent-holders of a mutant gene hold the rights to any future medical uses tied to its mutation—from drugs based on it to tools that might diagnose people who have it.

In the spring of '93 Hardy stumbled across a letter in the lab's fax machine that hit an already raw nerve. Sent by Mullan to a biotech firm in California, it led Hardy to believe that his colleague was on the verge of an agreement that would result in commercially available transgenic mice bearing the Swedish mutation. Hardy "couldn't countenance that," as he says today, and shipped off 300 vials of DNA, via FedEx, to a colleague in London for safekeeping. The missing DNA was quickly noticed, and the shipment halted and eventually returned to the Florida campus. Mullan today maintains that Hardy misconstrued his communication with the biotech firm and that it related only to work on chromosome 14, which Hardy and Mullan were putting their shoulder to.

While university officials attempted to resolve the dispute, Mullan changed labs within the department. Eventually, in four years' time, Hardy would accept an offer at the Mayo Clinic in Jacksonsville, on Florida's opposite coast. In hindsight, both scientists feel that when they arrived in America, certain external forces exacerbated a misunderstanding between them. "All this might not have happened, but we weren't very well organized when we came to America," notes Mullan.

By the start of 1994, in Building 149 we had gotten the suspect area on chromosome 14 down to around 10 million bases. Other groups appeared to be making equal progress. This was as far as my team could reliably narrow the region using standard positional cloning techniques. From there on, we looked for similar patterns of variants across our Alzheimer families that contrasted to those seen in the public at large.

Peter Hyslop and I daily exchanged E-mails about bits and pieces of our chromosome 14 work, many filled with extraneous chitchat. Hyslop was altogether quite cursory about his headway on the chromosome yet highly amusing about other parts of his life. After his wife gave birth to twins, he joked in one E-mail, for instance, that he was thinking of naming them "alpha" and "beta."

The more the field skimmed down the region, the more the competition stiffened, but for the most part remained amicable. After meetings, we downed beers together, and, in between other subjects, snippets of information would be shrewdly traded about where each team was positioned on the chromosome or which markers had been ruled out. There was a decent amount of reciprocity, but within limits—limits as stringent as those between Kennedy and Castro during the Bay of Pigs. The Florida team, meanwhile, was re-sorting themselves for the pursuit on chromosome 14. Mullan's lab was continuing the search on its own. John Hardy and Alison Goate had gone on to collaborate with SmithKline Beecham and the Institute for Genome Research. And Christine Van Broeckhoven, formerly partnered with Hardy and Mullan, had developed a liaison with the biotech company Innogenetics.

Raising all the more curiosity about the gene we were after, and whether it merged into the amyloid hypothesis, more new data spilled out about how Alzheimer's amyloid got generated.

It was known that the little A-beta peptide that broke free and glomped into brain amyloid came in varying lengths ranging from 39 to

43 amino acids. A-beta 40 was by far the most prominent length, but longer A-beta 42 appeared to be much more aggregable and therefore all the more treacherous. The new insight from Steve Younkin's lab was that the three earliest-found mutations on APP, which sat within a few bases of each other, lent to the APP protein's production of an increased percentage of A-beta 42 over A-beta 40. The Swedish mutation, which sat further away in the APP gene, previously had been found to precipitate an increase in both A-beta 40 and 42. In 1993, a classic paper by Peter Lansbury, an HMS biochemist currently at Brigham and Women's, had suggested that A-beta 42, in fact, might be amyloid's seed—the nidus around which shorter A-beta fibrils and other proteins clumped.

By the summer of 1994, many of us slogging along chromosome 14 had come to believe that *the* gene lay within a certain 3-million-base stretch. This region, spread out between the markers D14S61 and D14S289 on the long arm, was still large, but the gene was plausibly within striking range, which added a nerve-wracking intensity to the research. One more spin of the cage, it seemed, and someone would surely holler—for real, this time—"Bingo!"

"Every month or so, a new rumor circulated" that the gene had been found, remembers Wilma Wasco, who was helping me direct and coordinate our chromosome 14 pursuit. "Stories would start up and they'd snowball. We probably started a few ourselves without even knowing it. There was also the threat that the gene might be found by accident"—by someone looking for an entirely different gene.

One especially pesky rumor returned over and over again, each time in a slightly more believable and bothersome form. A Japanese postdoc, who worked in either Japan or New York, told someone who told someone else that, while investigating another disease gene, he or she had found a candidate gene for Alzheimer's on chromosome 14 and had given it to a prominent Alzheimer team that happened to work in the same building, and it turned out to be the real thing. We never entirely believed the Japanese postdoc rumor, but we always half-feared it. Whenever it circled back, we'd try to pick it apart as quickly as possible so it wouldn't hang over our heads.

"Nothing new on the old grapevine yet," John Hardy wrote me at one point, "but I'm asking every Japanese person I know if they know! Best wishes and keep those brown undies on!" (The last, in reference to the advantages of wearing brown in case one gets overexcited, was Hardy's bit of British humor.)

The search at this point became an educated guessing game. Which of the 100 or so genes that lay within the cordoned-off 3-million-base region was the offender? "Ideally," explains Wasco, "you could sequence the whole region, both in a normal person and someone with the disease, and compare their sequences to find the mutation. But sequencing that big a chunk was expensive, time-consuming, and an organizational nightmare"—much more so than now, when a genomics company could do the sequencing at a very fast clip. The widely used approach, instead, was to build a rational case for already-known genes in the area—what about their protein's function might apply to Alzheimer's destructive pathway? Those of interest we isolated from our families and laboriously sequenced, on the lookout for a mutation. Additionally, we inspected random pieces of novel genes, using information from homologous pieces in GenBank and our intuition to determine those worth pursuing.

During our analyses, a conspicuous pattern emerged. Among our twenty largest families, a certain 100,000-base-long segment in the region under focus statistically bore an association with the disease that was roughly three times higher than seen in the general population. Both in people with and without the disease, the area was such a cryptic, convoluted hall of mirrors, we took to calling it the "Bermuda Triangle." Its DNA was riddled with base-pair inversions, repeats, and other oddities, and it was easy to imagine that such genetic chaos might induce an Alzheimer defect. Taking a big leap of faith, Wasco and I decided to concentrate on this one area. I communicated this decision to Peter Hyslop, who said that while his Toronto group was also looking roughly in the same region, they were focusing on an area that lay millions of bases to the south. Our collaboration being murky, we never did establish a very black-and-white division of labor between us. Other groups similarly were carving out their own niches to survey.

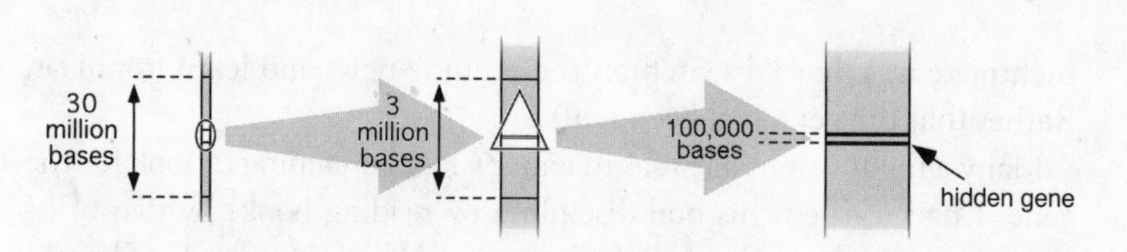

FIGURE 9.1 Chromosome 14 suspect region narrowed from 30 million bases to 3 million bases, then to the Bermuda Triangle. Illustration: Robert D. Moir

We settled into a routine in my lab of analyzing SPITS—or "slightly possible interesting transcripts," our term for RNA sequences that represented bits of genes from our Bermuda Triangle region that might contain a mutation. Before sequencing a gene, we reviewed hundreds of SPITS, singling out only those that GenBank's sequences and our own gut feeling suggested, if corrupted, might cause Alzheimer's. The danger was that we might mistakenly exclude from analysis a SPIT that actually came from the gene. "The big joke was, How will we know the gene when we find it? It wouldn't be waving a big red flag saying 'I'm the gene,'" recalls Wasco. Made daffy by all the SPITS confronting us, sometimes for fun we wound up one of those small yellow metal chickens that pecks away at a furious rate, placed it on a list of SPITS, and whatever SPIT its beak stopped at we gave serious consideration. At least for about a minute. Comic relief was always welcome.

Our command post's activity was reminiscent of General Eisenhower's on the eve of D day. The walls of the office I shared with Wasco were covered with genetic maps and SPIT printouts. Our multiple battle fronts included mapping DNA within the Bermuda Triangle, physically isolating DNA, and, when it seemed worthwhile, analyzing certain sequences for mutations—all this for the sake of cornering one tiny but devastating defect. Even though early-onset Alzheimer's is referred to as "rare," at any one time hundreds of thousands of people are trapped in its shadow.

Whenever I answered the phone, I held my breath, always half-expecting to hear that another team had *the* gene. My other ongoing nightmare was that I'd switch on the evening news and learn from Dan Rather that the gene had been nailed.

Many a night I went so far as to employ lucid dreaming to look for the gene. I had learned this odd discipline by reading books by two of its master practitioners, Stephen LaBerge and Carlos Castaneda. The objective is to be deliberately conscious while you're fast asleep and dreaming so that you can explore your dreamscape, essentially your subconscious, with the same awareness you possess when you're awake—even a heightened awareness, since you aren't distracted by normal stresses. When I was in this state and elected to, I'd ask whoever was present if they knew the mutant gene's identity. In one dream, a small gnarly man responded by holding out a box of Crayola crayons with its rainbow of colors. In another dream, a more anonymous figure held a prism under a bright light, evoking the same spectrum. Everyone in the lab thought I was crazy. Nonetheless, always glad for lighter moments, they tried to match these and other clues to our lab work. Perhaps the twice-dreamt rainbow spectrum referred to spectrin, a chromosome 14 gene associated with the cytoskeleton, Donna Romano suggested. There was instant excitement, until we remembered that we'd already excluded spectrin from our candidate region.

One November day in 1994, Wasco and I were in our office eating Building 149's vegetarian lunch special—a bean and rice casserole. Between bites, Wasco was scanning GenBank for information on an interesting SPIT of a gene we'd come across. Our Bermuda Triangle's 100,000-base stretch had yielded SPITS from an estimated five genes, but we hadn't deduced anything of interest about any of them. Suddenly Wasco let out a whoop. The SPIT she was researching bore a high degree of likeness to a GenBank gene that belonged to a gene family that made proteases—proteins that snip other proteins. Everyone was hoping to find such genes on chromosome 14 because they were such perfect candidates for the gene. A protease made overly active by a mutation might be interacting with APP's protein to release too much of its A-beta fragment.

It was a rousing discovery from another standpoint. Never before had this class of protease been seen in humans, nor in any mammal. The next species down from humans recorded to have it was the fluke, a parasitic flatworm. Jack-beans and rice also had it—exactly what we were eating for lunch that day! "I'd been running DNA sequences through GenBank month after month without any luck, so it was a thrill to finally see this great match," remembers Wasco. "It's what keeps you going, finding something like that."

We'd been excited when we'd wandered into the weird Bermuda Triangle stretch. But hitting upon a protease! We were so convinced that it might be *the* Alzheimer's gene that, taking another leap of faith, we opted to devote most of the lab to decoding its bases and examining them for mutations. We said nothing to the outside world about our find—not to Jim Gusella or other close MGH colleagues; not even to our techs, who set about sequencing the SPIT unaware of our high hopes for it.

As the weeks slipped by, to keep up with our techs' output, Wasco and I stayed in the lab later and later. It became somewhat of a custom to down a round of single malt for fortification before the chore of reading sequences and entering them into the computer. "Robert, the night janitor, became our pal," remembers Wasco, "and we began confiding in him about what we were up to, because we couldn't tell anyone else." One mention of the SPIT to another team, and it might either start a false rumor that we'd found the gene—and we didn't know that yet—or bring others to our vicinity on chromosome 14.

Outside, sails were reappearing on the Charles, the swans were back in the Public Garden's figure-eight lagoon, and Fenway Park was being gussied up for a new season of action. But inside the lab, frustratingly enough, our tests weren't registering so much as a suspicion of a mutation. I decided to mention our gene candidate at a talk at Washington University in St. Louis, especially to stave off the hungry rumor mill and categorically state we had something promising but not yet proven. Besides, at that point we wouldn't have minded a collaborator's help in locating the mutation. As I should have foreseen, my announcement backfired. The grapevine misconstrued the news, some people thinking we really had found an Alzheimer mutation—had captured the gene. Even Hyslop, hearing talk to that effect, E-mailed me to ask if we had it.

Toward the middle of May I got a call from Tsunao Saitoh, an Alzheimer researcher at the University of California, San Diego. He told me the gene had been found, but said he wasn't at liberty to mention by whom. I shrugged this off as either the rumor we ourselves had unwittingly started or the Japanese postdoc rumor circling round again. A day or two later, another colleague phoned with the same news. When he mentioned that the discoverer's paper had already been sent to *Nature*, I realized that what he was saying really might be true. He was surprised I didn't know, for the team leader was my collaborator, Peter Hyslop. I would soon learn that Hyslop had pulled the gene from the Bermuda Triangle, the very turf my lab had been frantically probing. Scooped in my own backyard! In fact, the mutant gene lay very close to the SPIT gene we'd been so busily analyzing. And all the while, although I knew Hyslop and coworkers were checking out the vicinity, I'd been under the impression that they mainly were occupied with a region many "miles" to the south.

Hyslop and I had made an agreement long before that if one of us happened to be the first to find the gene, or learned its identity from another source, neither of us would tell the other its identity without asking the others' permission. Because if the one being told was already on to it, it could corrupt the authenticity of his finding. But I never imagined that if one of us found it we'd keep it totally under wraps from the other right up to submitting a paper. The breakthrough could have been mentioned without giving away the gene's identity.

Right after hanging up from this last conversation, I called Hyslop for confirmation. He neither confirmed nor denied that he had the gene, but made some nonreply, which was as good as confirmation.

Both phone calls had taken place in Wilma's and my office, and, after hanging up with Hyslop, I stood there for a moment feeling a mix of emotions. On the one hand, I was very let down; we'd missed the gene by the narrowest margin. On the other hand, there was a sense of relief. The long search was finally over; another crucial piece of the puzzle was locked into place and we could get onto something new, something all the closer to a full picture of what caused Alzheimer's. As I'd recently told a reporter about the chromosome 14 race, "I can't wait for it to be over. I'd rather get back to enjoying the science."

With time, the race would blend into something much bigger. Researchers don't get the immediate gratification a physician receives by helping people every day. What we do get, in the long haul, is the gratification of knowing that through our own very occasional breakthroughs, and more often through the field's collective bent, we can positively affect humanity forever, and that's a large, very large, privilege. In the case of the chromosome 14 gene, now known as *presenilin 1*, the field's headlong charge had revealed, as would be seen, a gene whose scores of mutations cause early Alzheimer's in over one-quarter of a million people. Like APP, the new gene might impart a great deal about the disease's underlying mechanism in all 14 million people lost to the disease.

It was a late Friday afternoon, and as I went next door to our lab's administrative office, I could tell from the sound of chatter and laughter that our techs and postdocs had gathered to shoot the breeze and pop open a few beers, as they often did at the close of the week. I'd have to tell them and Wilma Wasco, my right hand, that it was over. They too would be disappointed at first—then relieved—and, with time, more and more proud to have been part of the effort.

Fran's symptoms broke down the denial that had separated her and her brothers and sisters from their mother's disease. They mourned just as if another death in the family had occurred. To the younger ones it was as if they were losing a second mother, Fran having taken over and helped raise them when Julia had become so sick. A dispiriting, troubling waiting game began. Who among them would be next? They questioned their small lapses when alone; they watched one another when together, trying to relieve the suspicions raised by every misplaced item, every irretrievable word, every vacant look. Some among them sought out counseling and what support groups existed. They began taking the A, B, C, and E vitamins, anti-inflammatories, blue-green algae, ginkgo, rosemary—anything they'd heard might keep the disease at bay. Malcolm cried a lot, and he prayed a lot. "God, may I find your will for me for another day."

— ten —

The 42 Nidus

What but design of darkness to appall?—
If design govern in a thing so small.
—Robert Frost, "Design"

On a June morning in 1995, after getting a cup of the Neurogenetics Unit's notoriously bad coffee, Wilma Wasco began typing out the letters corresponding to the 467-amino-acid sequence of the miscreant protein made by the gene that Peter Hyslop's lab had netted on chromosome 14. Curiosity had gotten the better of her. Would any lookalike sequences in GenBank's huge database help explain the protein's normal function in humans? Or how, if altered, it caused early-onset Alzheimer's? Even a like protein in a mole or an amoeba might say volumes about the human protein. And the question weighing most heavily, Could a GenBank homologue ultimately shed light on how this newly chased-down gene and protein interacted with APP to form A-beta and amyloid?

The week before, Wasco and I had traveled to the University of Toronto to consult with Hyslop about his gene finding. Inside the Centre for Research into Neurodegenerative Diseases, which Hyslop directed, you could feel the fevered intensity. People walking by us were so consumed, they seemed hypnotized. "It took us approximately six months to complete the work, from knowing we had the gene to having something that was ready for publication," today recounts molecular geneticist Robin Sherrington, Peter Hyslop's right-hand man on the project, who now works at Axys Pharmaceuticals. "Peter was totally dedicated, totally focused. For months he worked seven days a week. He wasn't one to go fishing with. He wanted to be the first person to clone this gene more than anyone else. It was the culmination of years of research, as it was for Rudy."

Even once the Toronto team hooked the mutant gene—first spied in Mass General's Canadian family—they'd kept fearing that it would appear in healthy individuals, which would have cancelled out their suspect. "Every time we expected [the gene] to curl up at the edges and blow away," Hyslop described to *U.S. News & World Report*. "It didn't."

From the get-go we called the new gene on chromosome 14 "presenilin" due to its association with presenile, early-onset Alzheimer's. In the six Alzheimer families Hyslop and coworkers analyzed, they initially detected five presenilin mutations. The gene, it would turn out, was awash in early-onset errors. By today's count, over eighty mutations have been identified in over 100 families of various ethnic origins, several having been unearthed in my lab. A defective presenilin not only explained the disease in the four MGH families Hyslop and I had been investigating for so long, but proved to be the problem in a Colombian clan gathered over the years by Kenneth Kosik, an HMS cell biologist at Brigham and Women's Hospital. To this day, this unfortunate family is by far the largest early-onset Alzheimer family on record, its known past and present, affected and unaffected members numbering close to 4,000. When Kosik first tracked down its kin in villages spread throughout the region of Antioquia, "they were very aware that what they had was an illness, not senility," recalls Kosik. "They called it *bobo*—'stupid' in Spanish. Some were very superstitious; they thought they could catch the disease by touching a certain tree, though they didn't know which tree."

All the time it was a tree—their family tree. So comparable to the large Venezuelan family burdened by Huntington's disease, its branches had been prodigiously spreading due to the tendency for many Latin Americans to bear numerous children.

Because of our collaborating role and inclusion as coauthors on Hyslop's forthcoming report, Hyslop had given Wasco and me the gene's sequence so we could get a head start in examining it, with the explicit agreement that we'd keep it under wraps until the new gene's account aired a few weeks later in *Nature*. As Hyslop conveyed to us and his paper would report, his group indeed had sighted a presenilin-like protein in GenBank. It belonged to C. *elegans*, the tiny dirt worm, and played a role in transporting other proteins vital to the formation of the worm's sperm, of all things. As Dennis Selkoe would remark to the press, "One would have to say that this is not what one would have expected"—a worm protein related to worm sperm that structurally resembled a human protein that, when mutated, caused a brain disease!

Wasco had decided to use the new gene to do her own GenBank search just in case another presenilin-like gene or protein had been passed over. "I was grumbling," remembers Wasco. "When you transcribe a long sequence, it's easy to lose your place, and to save time I'd wanted Rudy to read the letters off while I typed them down." But besieged by phone calls, I couldn't help her out. Having caught wind of Hyslop's coup, colleagues were calling in droves. How had my lab missed finding the chromosome 14 mutant when we'd practically been sitting on it? I had no excuses, but never again would I put all the lab's eggs into one basket. If we hadn't been so hoodwinked by our protease candidate, we might have looked more closely at nearby genes and snagged the real culprit. Those calling were especially curious to learn if I had any information about Hyslop's gene. Beholden to confidentiality, I couldn't help them out.

So Wasco muddled by on her own, her cup of dismal-tasting coffee her only companion. (Jim Gusella had read somewhere that a natural chemical in coffee was also found in skunk spray, albeit in a much more concentrated form. The unit's coffee that day was an especially stern reminder of this.) Midmorning, finally finishing her task, she entered the presenilin's amino-acid sequence into GenBank and no sooner had done so when up on her computer screen popped not only its look-alike in worm, which Hyslop's lab had spotted, but also a piece of an unidentified human protein. It bore a striking 80 percent likeness to the new Alzheimer protein, so it was undeniably interesting. Hyslop hadn't mentioned this GenBank sequence to me, which seemed a little strange. Stranger still, it would not appear in his forthcoming *Nature* report.

As it happened, James Sikela at the University of Colorado Health Sciences Center had found this protein snippet in 1993. Its genomic location was unknown. Such orphan bits of unidentified genes are called ESTs—or expressed sequence tags. Sikela as well as Craig Venter at the NIH were the first to submit them to GenBank, which is now flooded with them.

"The moment I saw it," says Wasco, "I knew it had be a protein belonging to a sister gene of presenilin." Thus, very possibly, a protein associated with yet another Alzheimer gene! "The very next moment, something else flashed through my mind—maybe it was the Volga German gene."

The genetic error responsible for early-onset Alzheimer's in a small number of Volga German families had been eluding Jerry Schellenberg and his University of Washington crew for nearly a decade. The population is so-named because, while originally from Germany, in the 1760s they'd received the "Great Call" by Russia's empress, German-born Catherine the Great, who invited them to settle on the fertile plains surrounding Russia's Volga River. In time, many moved on to the United States. Schellenberg's collaborator Tom Bird had collected nine of these families, each blatantly prone to Alzheimer's as seen from their lineage, but the identity of their mutated gene had become more and more baffling. Their APP gene on chromosome 21 was clean. No Alzheimer mutations there. And even before the chromosome 14 gene had been isolated, its genes also had been ruled out due to the lack of any incriminating linkage to 14.

Wasco showed me the GenBank EST, and my mind similarly leapt to the Volga Germans. They were the only sizable collected early-onset Alzheimer kinship whose mutation hadn't yet been found. A week or two before, Schellenberg had confided to me that his Seattle group finally had traced the Volga Germans' defect to a chromosome: chromosome 1. He'd even submitted his lab's linkage report to *Science*. Yet chromosome 1 being the longest human chromosome, it might take months, even years, before his lab learned which of its thousands of genes was marred. And here, on the computer screen, the very gene might be in front of our noses.

At Mass General we had a Volga German brain in the freezer sent some time ago by Bird. Using that tissue, we could have set out on our own to isolate the complete presenilin look-alike gene found by Wasco and thereby discover whether it was the answer to the Volga German's disorder. But I wasn't keen on scooping a good friend who'd been tracking the Volga German's DNA error for many years. If Schellenberg had the correct chromosome, and we really had hold of the right gene, a partnership was the most expedient, collegial way of turning one and one into two.

Calling Schellenberg, I poured out the news: We had a great Alzheimer candidate for his families, and our labs should consider a formal collaboration. But I couldn't divulge the identity of the GenBank gene; it was too obvious a look-alike to Hyslop's isolated presenilin, which I had to stay mum about until its publication. For now my lab could fish out the GenBank gene and, by comparing its normal version to the Volga German sample in our freezer, start looking for mutations. And at least we could send Schellenberg DNA primers corresponding to the novel gene (without their sequence information) so he could check to see if the gene actually sat on chromosome 1. Schellenberg agreed to team up, but he sounded a mite leery. He was seasoned enough not to think "eureka" before having tangible evidence in hand. "I felt the chances were somewhat remote, but worth pursuing," recalls Schellenberg.

During the next few days, Donna Romano fashioned a pair of primers corresponding to the GenBank EST piece and isolated its gene, and, using the same primer pair, Warren Pettingell, another senior technician in my lab, proceeded to look for mutations in just the little EST section of the gene. The odds were that it would take many sets of primers spanning the entire gene before we hit the mutation. Schellenberg and I were in touch again. A warming urgency had replaced the caution in his voice. The probes we'd sent him confirmed that the new Alzheimer candidate lay not only on chromosome 1, but exactly within the region his team had staked out.

On June 29, 1995, to widespread fanfare, Peter Hyslop's chromosome 14 gene discovery ran in *Nature*. "The *Nature* paper ends one of the decade's hardest-fought competitions in biomedicine," described *U.S. News & World Report* in a lengthy article on the chromosome 14 race that depicted Alzheimer genetics as "one of the most cutthroat areas of research." The cliché was pretty accurate. What was it about our field? Were our egos sadly all the larger because of the large, terrible disease we were pitted against? Was it the potentially sizable professional and financial reward of getting to the other side of Alzheimer's? Both, perhaps, and more. For fate seemed to have arbitrarily thrown together more than a handful of particularly intense personalities.

As John Hardy once surmised about the field in an E-mail to me, "[It's] a mixture of idealism, selfishness, generosity, greed, fun, anger, sex, drugs, and rock 'n' roll." The less harsh way of looking at it was that the hot sparks of conflict, singeing so many of our butts, were making us charge forward as fast as we could go.

The gene breakthrough on chromosome 14 was major enough that *USA Today*, following the lead of ABC News's more limited coverage, broke the story a day before the gene's official report in *Nature*, thus overstepping the embargo period some science journals request of reporters so that a journal's contents reach subscribers before the media exhales any major news to the rest of the world. This didn't sit well with *Nature*. A week later it announced it was dropping *USA Today* from its press distribution list.

I came across the *USA Today* article while sitting in the waiting room of a spa in Calistoga. A recent recipient of a Pew Scholar award, I'd arrived in California for a Pew organizational meeting, and, still reeling from having lost the chromosome 14 race by such a slim margin, I'd taken an extra day to tour Napa Valley's wineries and submit to a therapeutic overhaul—steam, mud bath, Jacuzzi, Swedish massage, the works—hoping it would be restorative. The spa visit worked wonders, but the very next day my equanimity was shot through once more, nearly quite literally. As I emerged from a restaurant in San Francisco's Chinatown, firecrackers seemed to be exploding in short bursts all around me. When a pedestrian a few feet away dropped to the ground, the bottom of his T-shirt red with blood, I tried to help him, and meanwhile watched as several other nearby pedestrians, dropping belly-down to protect themselves, got hit by bullets ricocheting off the pavement. The cross fire, which was being exchanged by two rival street gangs, seriously injured seven passersby. Only sheer luck had kept me safe. Two years later, I'd be subpoenaed as a trial witness for the case. What I've come to take for granted is that my life seems to constantly play out this way. It's as though I have a sixth sense for wandering into trouble and find myself constantly picking my way through land mines.

Once the presenilin gene's sequence was in print, Schellenberg and coworkers used it to obtain its look-alike EST in GenBank, just as our Boston team had done. Their collaborator Darwin Molecular then began sequencing the EST's complete gene. Darwin—a biotech company north of Seattle that was underwriting a gene discovery program for Schellenberg's lab—had recently been launched with backing from Microsoft's Bill Gates and Paul Allen. To be certain, science's interpretation of DNA couldn't be happening so rapidly were it not for the fact that the computer's binary code, and its ability to process and store voluminous data, was increasingly helping to make sense of DNA's quaternary code of bases—A, T, C, and G.

When Schellenberg and I spoke again on the Fourth of July, a Monday holiday, Schellenberg sounded very urgent, very aware. Something was definitely up. I asked if Darwin had finished sequencing the suspect gene in his Volga German families. It had, he replied. How about you? We were still in the process, I told him, but Pettingell had sequenced the gene's little EST segment in our Volga German sample. Moreover, Pettingell had indicated he'd noticed something, but having been away in California I hadn't had the chance to look at his autorad results or notebook entries. They were sitting on a corner of my desk, and I began reviewing them, meanwhile putting it to Schellenberg fairly bluntly: So had they found something? Well, maybe, said Schellenberg. A mutation? I asked. I think so, he replied.

Here we reached a fairly sticky juncture. If he told me what the mutation was, and I checked the autorads and told him that Pettingell had come across the same error, for all he knew I was just repeating back his information. In our line of work, even among hombres any amount of suspicion can worm its way in, and that's how friendships can wither and die. I suggested that if he told me their finding, and Pettingell had spotted the same thing, to confirm my lab's independent finding I would FedEx him copies of our raw data that day. Schellenberg disclosed the mutation on the new gene. It was an A-to-T base change, and it appeared in seven of his nine Volga German families. The alteration mucked up the protein by inserting an I (isoleucine) in place of an N (asparagine). Pettingell's data indicated the same alteration in our brain sample, and I conveyed this to Schellenberg.

Later that day, after several phone calls, Wilma and I tracked down the one FedEx office in Boston open on the holiday and sent Schellenberg copies of our filmed evidence, after which we celebrated this latest advancement by plunking ourselves down at Small Planet's bar in Copley Square for a round of beers. Less than a week had passed since the chromosome 14 gene's publication, and now a *fourth* Alzheimer gene had been nabbed. Scientific progress hardly ever happens so fast.

This latest early-onset gene would go by the name of *presenilin 2*, or PS2. Posthaste, Schellenberg and I began writing up our formal report, a finding never sealed until it's in print. With presenilin 1's sequence now common knowledge, other groups could enter GenBank and, just as we had, quickly discover its sister gene. As I later learned, several other teams were in the process of doing just that. In fact, John Hardy had called Schellenberg to suggest a collaborative effort on chromosome 1. As Schellenberg's and my gene report was in the oven—not yet finished or published—and it was too early for Schellenberg to share the finding, he had immediately switched subjects to the first thing that entered his head, which happened to be his pet turtles.

In the meantime I had called Peter Hyslop in Toronto to let him know just what our GenBank search had yielded—the long-sought Volga German gene. I asked him why, in his *Nature* paper, he did not divulge the GenBank human homologue, which he should have come across while preparing his presenilin paper. It soon became clear: Even though Hyslop hadn't mentioned the homologue in his paper, his team had pursued it, isolated it, and was preparing its report.

On July 12, 1995, George Glenner died, taking with him the knowledge, at least, that one more Alzheimer gene—presenilin 1—had been found; one more appreciable clue to help make sense of "the amyloid story."

The amyloid obstructing Glenner's heart had kept him pretty much bedridden the last six months, but it never dimmed his interest in how things were proceeding in the field, at the Glenners' day-care facilities, or in his own lab, where, wheelchair-bound, he returned whenever he had the strength. Nor did his illness break his humor. Uncertain of Glenner's lucidity, one Saturday toward the end his doctor had asked, "Do you know where you are?" Glenner, his eyes closed and his head sunk back in a pillow, replied at once. "Nordstrom's." Recalling the scene, Joy Glenner can't help but laugh. "Doctor Blanchard thought George had really lost it. But he hadn't at all," she explains. "He and I had spent so many Saturdays shopping together at Nordstrom's, he was just being funny."

Glenner died believing he'd found beta-secretase, one of the two enzymes responsible for clipping free the A-beta peptide from its larger protein and therefore an accomplice in amyloid aggregation. Since 1992 he'd been working to further describe and verify it. "Here it is," he said that year to a *San Diego Union-Tribune* reporter, exhibiting a slight mark on a gel and pleased to share evidence of a molecule whose identity he felt could lead to an amyloid-inhibiting Alzheimer drug. "If you could prevent the clipping of [A-beta], in theory you could prevent amyloid fiber formation." After Glenner's death, those in his lab would continue to try to confirm that his enzyme was the real thing.

Glenner, nonetheless, had left behind one certain tangible—the A-beta peptide. This achievement was leading so many scientists forward, that he'd posthumously be called the Pied Piper. As Robert Katzman, rising above their at-odds relationship, remarked to a reporter in late '95, Glenner "opened up a wild fire and the basis of 80 percent of current Alzheimer research." Along with Glenner's feats at the molecular level, in 1994 he and Joy had launched a third day-care center in the San Diego area, their facilities by then renamed the George G. Glenner Alzheimer's Family Centers. They also had established the Glenner School of Dementia Care, a certified vocational school with a welfare-to-work training program for nurses' aides specializing in dementia.

To this day there's no explaining how Glenner happened to have contracted an amyloid disease. None within this class of diseases are known to be infectious, with the exception of the prion diseases. Robert Terry at UCSD notes that since Glenner's death he's heard of a few other amyloid investigators who also have succumbed to cardiac amyloidosis. Could their close work with amyloid be to blame? "I have no idea," says Terry. "But it's definitely worthy of investigation."

Glenner probably would have been amazed, as were we all, that immediately on the heels of a second early-onset Alzheimer gene, a sister gene—presenilin 2—had tumbled into view. The August 18, 1995, issue of *Science* carried both Jerry Schellenberg's initial linkage to chromosome 1 as well as his and my lab's joint report of its faulty gene. Submitted only a month earlier, the latter paper underwent about as quick a submission-to-publication turnaround as a journal is able to do. As Wilma Wasco notes, "Editors want to win too," especially when it comes to encasing a newly unveiled disease gene in print.

Scientific and public interest in disease genes, stoked by their quickening discovery, were at an all-time high. The consequence of a broken bone was probably grasped to some degree as far back as when our distant ancestors swung from the trees (and occasionally missed a branch). Yet here it was late in the twentieth century and only now was it becoming fully apparent—so much more so than in our parents' generation—to what extent altered genes in the depths of cell nuclei contributed to thousands of disorders. And not only heart disease and a range of cancers, but also conditions few would have guessed had heritable aspects. Schizophrenia and other psychiatric conditions, autism, forms of epilepsy and deafness, cataracts, asthma, sperm and egg deficiencies, obesity, reading disorders, high blood pressure, on and on. As would be seen, gene variants even can contribute to anxiety as well as sensitivity to pain. What a sobering view into the warts-and-all machinery inside all of us.

Momentarily, the presenilin 2 gene rode the news waves. "One Alzheimer's Gene Leads to Another," featured *Science News*. "Third Gene for Inherited Alzheimer's Disease Found," blazoned the *L.A. Times*. Peter Hyslop's independent report of presenilin 2 followed in two weeks in *Nature*. A third PS2 account—by Huntington Potter's Harvard Med School lab—appeared in *Proceedings of The National Academy of Sciences* some months later. Back in 1988, when the field was frantically hunting for mutations on the APP gene, I'd attended an international meeting in Osaka where many a human gene jock was present and where I'd spent a good deal of time defending the issue that at least some forms of Alzheimer's had a genetic basis. Now that four genes underlying its dementia had been isolated—APP, PS1, PS2, and APOE—those days were definitely over.

It was soon established that presenilin 2 explained very few cases of early Alzheimer's. The presenilin 1 gene, on the other hand, accounted for so many that after the dust of the recent discoveries settled, the field began to realize just how far it had come in identifying the disorder's genetic seeds in people under sixty. As it stands today, we believe that mutations in the trio of genes—APP, PS1, and PS2—account for roughly 40 percent of early cases, with PS1 behind the lion's share. PS1 mutations also cause the disease's youngest fatalities. In extremely rare and brutal instances, victims can be in their twenties when dementia becomes noticeable.

These three early-onset genes were but the tip of the iceberg. Of that the field was very much aware. Altogether, it would be determined, they cause less than *2 percent* of Alzheimer's total cases, young and old. But many of us were of the opinion that their relentless mutations could be invaluable in elucidating the disease's step-by-step unwinding at whatever age. Anyone born with only one of these mutations always got the disease, and there was a marked generation-to-generation regularity in the numbers of offspring afflicted, neither of which appeared true in late-onset cases. It seemed that such virulent pathology surely could provide clues about the disease's late form.

For Baptists, there was a tremendous amount riding on the discovered presenilins. We expected to find—in keeping with the amyloid hypothesis and the notion that A-beta was the disease's lethal barb—that the mutant presenilin proteins worked in cahoots with the APP-amyloid protein, influencing APP's output of A-beta and amyloid in the brain. If they didn't, we Baptists would be up a creek without a paddle. Our hypothesis might be shattered, and we'd have to completely rethink the presenilins' involvement in the disease.

Yet even while we were searching for the PS1 gene on chromosome 14, before it had been identified, there'd been signs that this gene—and, once it was caught sight of, its sister gene PS2—would play into our hand.

Earlier in 1995, contacting Dennis Selkoe, Jerry Schellenberg, and me, Steve Younkin at Case Reserve had more or less put down a wager that had to do with the families we'd linked to chromosome 14. Younkin had previously shown that mutations on the APP gene led to increased levels of A-beta 42, the longer and fiercely aggregable form of the A-beta fragment. Like many of us, he was of the mind that Alzheimer mutations in yet-to-be-found genes somehow must interact with APP to similarly elevate A-beta 42. "If you provide me with the fibroblast lines"—the skin cells—"of those families whose disease-afflicted members you've linked to chromosome 14, and if you withhold who in each family has the disease and who doesn't, I believe I'll be able to tell you just that."

His logic was impeccable. If A-beta 42 was central to Alzheimer's pathology, cells of patients taken from most anywhere in the body should be churning out more A-beta 42 than cells taken from their healthy siblings. And that's precisely what Younkin's lab observed in our chromosome 14–linked families: a roughly twofold increase of A-beta 42 relative to A-beta 40 in patients versus normals. Younkin announced this seminal finding later in '95, and it was subsequently published in *Nature Medicine*.

The impression was mounting that A-beta 42 could be a truly deplorable peptide. When clipped from its APP protein by the two enigmatic secretases, it quickly self-aggregated, became extremely subversive and insoluble, and progressively fouled things up by becoming the seed that A-beta 40 and other proteins glomped onto, forming a fattening plaque. Some of this was still hypothetical but believable enough. To me, A-beta 42 seemed about as complexly evil as Hannibal Lecter. I find I invariably anthropomorphize molecules this way, for it helps to interact with them and get a sense of what they might be up to. You just hope that in regarding a peptide or an enzyme as bad, it doesn't keep you from perceiving, one day, that it actually might be good. In the back of my mind I had to wonder whether A-beta wasn't *both* bad and good, somehow. There had to be a reason that healthy cells routinely spit out soluble forms of this peptide all over the body.

Leaving it to others to confirm that mutant presenilin boosted levels of A-beta 42 in the brain, my lab instead moved on to exploring the normal, everyday characteristics of the presenilin protein. Somewhere in the cell it had to be talking to the APP protein, if indeed its errant form prodded APP into overproducing A-beta. To determine whether the two proteins were indeed interacting, several questions needed answering. Which cells in the body processed the presenilins? Where did they hang out in the cell? What tasks did they normally carry out? Ours was hardly the only lab so occupied. "When the presenilins were first found, it was as if a cookie had been dropped and thousands of ants immediately crawled all over it," describes Tae-Wan Kim, a cell biologist in my lab.

My labmates Wilma Wasco, Dora Kovacs, and Kim secured crucial data. The presenilin genes, they detected, are highly conserved in most every organism higher than yeast and bacteria; evolution hasn't forfeited them, indicating their importance for an organism's survival. Most human cells utilize their proteins, particularly neurons. Even more telling, their proteins wind up waving in and out of the membranes of two organelles inside the cell, the endoplasmic reticulum and Golgi complex. The field was so hungry for any crumbs about where the presenilins might encounter APP in the cell that this crucial insight by Kovacs and Wasco, published in *Nature Medicine*, became one of the field's most cited papers.

Entrusting the lab's wet work to a circle of skilled recruits, I was at the bench less often, my time instead given over to running a lab. These responsibilities would increase in 1996 when my lab evolved into Mass General's Genetics and Aging Unit. Having followed the same road as Jim Gusella—skipping a postdoc and going directly from graduate student to lab head—I was in the process of gaining even more respect for how Gusella had broken in a lab, then a unit. I often wandered down the hall to get his brotherly advice on one oblique problem or another. That's how I'd come to view him—a big brother who always had the answers. The lab's senior scientists—Wilma Wasco and Ashley Bush—and I were trying to create a workplace that could do a little of everything—a little molecular genetics, a little cell biology, a little protein chemistry. This gave us lateral leeway. Around us, commercial and academic genomic enterprises were loading up on computer-driven sequencers and other automated fare, each machine as costly as a Lamborghini. This type of lab we couldn't be; rooms crowded with purring, fast-track equipment were beyond the reaches of a typical lab sustained by grant funds. But we could commission out sequencing and other brute-force jobs, while concentrating on the more elegant, interesting work of revealing and characterizing gene candidates.

In the midst of managing funds, or reviewing colleagues' pending reports and grants, or finding the best deal on lab equipment, or putting out fires—be they scientific or personal—I was as engrossed as ever in trying to figure out the mystery of Alzheimer's and spent as much time as possible problem-solving with my teammates. The clues pouring out of hundreds of laboratories, some fitting together, were making the entire puzzle seem solvable, even as it was becoming incredibly intricate.

My approach to science—then as now—was to balance intuition with careful scrutiny of the published data. This had as much in common with my gleanings from Taoism and its uncluttered "Way of Things" as with the extemporaneous Keith Jarrett–like piano style I worked on when time allowed. Since high school I'd immersed myself in Buddhism, Christianity, Sufism, Taoism, and other ancient observances, especially for their interpretations of the metaphysical world, the things we can't see, and how things came to be. To my way of thinking, Taoism and its practice of not intervening with the natural course of events for the sake of a pure outcome is the path toward life's master progenitor, its invisible perpetuator. I attempted to bring this unfettered response to the piano, freeing my mind so that my fingers would find a progression of notes and chords unadulterated by thought. Letting the music play through you, is how I regard it.

This approach could be extended to science as well. By studying as many reports as possible, when deciding upon a course of action I tried to trust that the subconscious truth inside me would emerge and carry the science in the right direction. "Try not to interfere with what you're doing," I tell those in the lab today. This would seem to be at odds with what science is all about—an attempt to intervene based on intellect. But it remains my belief that you must try to be merely a conduit for truth and not get in its way.

Not long ago, the actress René Russo asked Jerry Schellenberg and me about our belief in God. He and I were at an evening function in Santa Monica for the Cure Autism Now Foundation, which Russo was also attending, and the three of us ended up dining together. I definitely had all the symptoms of being starstruck, although Schellenberg—a dad whose movie selections are mostly confined to Disney films—didn't altogether know who Russo was. "Do you pray to God for scientific success for your experiments?" she asked us in a serious vein. With pure dash and no offensiveness, Schellenberg wondered aloud, "If I did, does that mean I'd have to make him a coauthor? Because if so, I'd have to think it over." She laughed, mentioning she'd have to give that some thought.

The activity of the presenilin protein would prove to be highly cryptic. Even today, many labs continue to scrutinize exactly where it and the APP protein rendezvous in the cell, and how this meeting might cause APP to generate more A-beta 42. I'll save you all the scientific twists and turns and leap to the most popular current theory. It could be that the presenilin protein escorts other proteins around the cell, and thus chaperones APP either to its workstation on the cell's outer membrane or to cellular entities that degrade proteins. During this brief encounter, presenilin might place APP in a spot that allows the gamma-secretase—a protease—to clip APP, liberating A-beta and generating amyloid. Should presenilin be mutated, all the more liberated A-beta 42 and amyloid might result. (It's believed that beta-secretase makes its cut in APP first, followed by the gamma-secretase's cut, which completely releases A-beta from APP.)

Steve Younkin's theory that a defective presenilin correlates with more A-beta 42 in the brain was proven airtight by numerous investigators. The most convincing verification came from transgenic mice. Since their disappointing debut in the early '90s, these potential mimics of Alzheimer's pathology had made significant gains. In 1995, renewing the field's optimism that mice could be important vehicles for drug-testing, Dora Games, a pharmacologist at Athena Neurosciences, presented the first mouse "that really shook the world," as Penn pathologist John Trojanowski describes it. The insertion of mutant APP into mice "produced a mouse with a headful of plaques. It was a sensational discovery." At long last we had living proof that APP mutations directly led to increased A-beta 42 and amyloid buildup. The older the mouse, the more A-beta 42 littered its brain.

The Athena mouse model originally had been engineered by scientists at the biotech company TSI, Inc. Apparently, TSI didn't realize how earthshaking its mouse line was, for Exemplar—a division of TSI—put the model up for sale. Athena, upon close scrutiny of the mouse, saw that it was both greatly overexpressing A-beta and producing an Alzheimer-like pathology and immediately bought out Exemplar.

Yet grumbles could be heard. Athena, a private company, was making precious few of its exceptional mice available to other research groups. Supported by the drug giant Eli Lilly, the Athena/Games' mouse apparently was too essential to the company's drug-testing pursuits to freely share with the competition. Working off public funds, those of us in academic science might have been reprimanded for not having passed around such a superlative mouse. By the summer of '96, the model, which had fifty-some Athena researchers perfecting it, was the object of a considerable clash between corporate and academic thinking. "I heard people attack Dora" for not sharing the model, "as if she was personally responsible," remembers one researcher. In time, Athena would send their mouse line to other labs—but only male mice sans gonads, which guaranteed that the line couldn't be bred.

Scientists in other establishments weren't to be held back. It wasn't long before other plaque-bearing mouse models were being successfully designed. Among them was an especially amyloid-prone mouse made by a team headed by Karen Hsiao at the University of Minnesota. It contained a defective APP gene altered by the Swedish mutation. Hsiao made the model available to nonprofit scientists, with the stipulation that if their institutions profited from experiments with the mice, royalties would be paid to the Mayo Clinic, which had licensed the mouse from Hsiao and Minnesota. Now that the presenilin genes had been captured, in quick succession their mutations also were incorporated into the rodent genome. Transgenic mice made by two Karens—Karen Hsiao as well as Karen Duff, now at NYU's Nathan Kline Institute—would come to represent the industry standard. "K2 mice," some call them. Crossing APP strains with presenilin strains, K2 mice exhibit especially robust brain plaques when mice, which on average live for twenty-two or so months, are only three months old. Because they produce amyloid so rapidly, these mice substantially shorten the time it takes to test a drug.

As tremendously encouraging as the transgenic scene was by 1996, the field didn't yet have a mouse that roared—a complete model of Alzheimer pathology that might be all the more useful for evaluating drugs. Engineered mice showed little sign of the widespread neuronal loss seen in human brains. And no models developed neurofibrillary tangles, although in some cases tau—the tangle protein—appeared to be altered. Since tangles were hardly ever seen in animals, the mouse brain simply might not be able to make them. Finally, when plaque-replete mice underwent water maze tasks and other tests, they showed little sign of cognitive decline, at least not to an extent that could be considered significant.

Each better mouse model, nonetheless, provided more proof of the pudding: Alzheimer's early-onset mutations increased the production of the A-beta 42 peptide, which seeded the formation of all the more amyloid deposits. While brain levels of A-beta 42 are minor compared to shorter forms of A-beta, A-beta 42 was looking highly dangerous: It was vastly more adhesive and quick to aggregate. As I wrote in *Nature Medicine*'s News & Views section in early '96: "In Douglas Adams's *Hitchhiker's Guide to the Galaxy*, the answer to life, the universe, and everything is simply '42.'" And so it appeared in Alzheimer's disease.

My article prefaced a study in *Nature Medicine* that further implicated the "42" peptide. Researchers at Case Western Reserve observed that a

Down syndrome fetus bore evidence that A-beta 42 peptides drove plaque formation beginning as early as twenty-one gestational weeks! The authors believed that A-beta 42 wasn't simply being overproduced by cells, but that the brain failed to clear it. This observation would become an important criterion for hunting down still other genes, notably those involved in the disease's late arrival.

From Fran: "At first I didn't know what was wrong with me, although I suspected. It's not like forgetting; it's different. It's just gone; the thought isn't there. It might come back if someone reminds me, but not necessarily. . . . When I was growing up, I discovered my love of carpentry and always had a screwdriver or a hammer in one hand or another—I used to be ambidextrous. I could bowl with one hand, then do the same with the other. But since the illness, I no longer am. . . . I've gone from reading novels, to *Sports Illustrated*, to *People* magazine, but now I've pretty much given up on reading. I try not to stay totally frustrated; otherwise I wouldn't have a chance to laugh and love. The worst pain I feel is over my children and husband. I don't want them to feel this pain that turns life upside down. I hate that my children will go through this awful disease. And who will be a grandmom for my grandchildren?"

— *eleven* —

Untangling a Cascade

The swiftly rising curve is not smooth.
—Buckminster Fuller, "Utopia or Oblivion"

Science pushed on. The sightings of the two latest early-onset genes—the presenilins—left dangling the possibility that maybe, just maybe, certain of their mutations also lent to late Alzheimer's, the variety that threw so many older people into fatal confusion, survivors who having weathered life's ups and downs would seem to deserve a more humane ending.

The Duke team's cornering of APOE, the only gene yet linked to this version, had brought alive the quandary over other late-acting genes. How many were there? On which chromosomes did they reside? As significant as the correlation between APOE-4 and the disease's late strike, people with this risk variant didn't necessarily fall victim, while many others without it did. The theory flourished that milling around in the population were several or even a dozen Alzheimer susceptibility genes. Each either acted alone or cocontributed to the disease's common assault. Other later-life genetic diseases were suspected of similarly stemming from multiple off-kilter genes. In my own mind's eye I envisioned some heavy-hitting gene variants, some weaker. Their polymorphisms, it would seem, were as commonly occurring in the human genome as the number of people who suffered their effects.

Genetic and epidemiologic studies that suggested an inherited tendency in a substantial portion of late-onset cases had been building for some time. But the dogma insisting that late Alzheimer's was due to environmental stressors and/or aging was so entrenched that the notion that many older cases might, in fact, be tied to inherited polymorphisms had only begun receiving wider attention since the early 1990s. In February '91 the *New York Times* had referred to it as a "dramatic switch" in understanding.

But for those who had been meticulously compiling the evidence for over a decade, it was hardly an epiphany. "Beginning in the early 1980s, my own work and that of other investigators showed that individuals who had a close relative with either late or early-onset Alzheimer's themselves faced an increased occurrence of three- to fourfold," recounts John Breitner, a psychiatrist and chairman of the Department of Mental Hygiene at Johns Hopkins University. "It was clear that even late-onset Alzheimer's ran in families. But even by the late 1980s some leading authorities dismissed this notion with the blanket statement that only 5 to 10 percent of Alzheimer's was inherited."

Among the older school, a chief criterion for an inherited illness was three or more close relatives with the disease. This strongly familial pattern, which usually was blatantly present in families with younger cases of Alzheimer's, wasn't nearly as observable in families in which older cases presented themselves in a checkered way, making some researchers stick to their guns that the disease's late arrival had nothing whatsoever to do with genetics.

Yet in the 1990s, studies of older twins were demonstrating heredity's noticeable hand in the disease's late strike. Often both twins got Alzheimer's; therefore genes must play a role. Two major studies—one from Sweden and one from Norway—suggested that two-thirds to three-quarters of the population's susceptibility to Alzheimer's could be traced to genes. The curious thing about the late-onset mechanism was that while it accounted for so many cases, within a family it lacked the *frequency of occurrence* seen in early onset, plus bore an environmental twist. In examining identical twins, Breitner and other researchers noted that the disease could emerge in one twin *a decade or more* after the other twin showed symptoms. Things in our surroundings, it would seem, could hasten or delay the disease's inherited mechanism. "Most of what we know about environmental influences is that they appear to modify the risk of an underlying genetic predisposition," notes Breitner.

Cancer genetics was providing a valuable parallel. Here, too, lay the suspicion that the vast, unsolved mountain of sporadic cases often were tied to multiple susceptibility genes that could be influenced by outside elements. Here, too, geneticists had found that a relatively small percentage—an estimated 5 percent—were solely and directly caused by a single genetic defect at an early age.

The presenilins, it was soon decided, had no major bearing on Alzheimer's onset after age sixty. The time having come for the field to delve more deeply into the late-onset wilderness, a major shift in strategy occurred. Clearly there were families whose early-onset pattern—particularly in the fifty-to-sixty-year range—remained unexplained, but by and large the era of searching out early-onset genes was ending.

Many a lab instead pursued two separate goals, both of which became the focus of my Mass General unit. The first was simply to determine more about the modus operandi of the early-onset mutations and the cascade of pathology they set in motion. Troubleshooting the early-onset mechanism might instruct the design of drugs for both forms of Alzheimer's. A second goal was to try to pin down other late-onset risk genes in addition to APOE. "The more obvious genetic aberrations—the big diamonds connected to early-onset Alzheimer's—were the ones found first. That's been true of other diseases, too," notes Deborah Blacker, a Mass General/HMS psychiatrist and epidemiologist. "Now we had to look for many small diamonds in a heap of shards of broken glass."

Fortunately for my Genetics and Aging Unit, a unique collaborative effort presented itself. Neuropsychologist Marilyn Albert, who was also based in Building 149, and coworkers at Harvard Medical School had been chosen by the National Institute of Mental Health (NIMH) to participate in the Alzheimer's Human Genetics Initiative, a project aimed at combing the genome primarily for more late-onset perpetrators. Also participating in this endeavor were researchers at Johns Hopkins and the University of Alabama at Birmingham.

The initiative's first step was to collect the DNAs of as many affected families as possible. By the summer of '96, over 300 families had been tracked down, most laced with the disease's late form. Joining forces, my crew and Albert's, which included Deborah Blacker, prepared for step two—the screening of DNA markers in these families, which with any luck would guide us to rogue genes.

Hundreds of labs planetwide were similarly consumed, although methods varied from lab to lab. Having encountered the same telltale evidence, everyone agreed that Alzheimer's late-onset mechanism looked to be a far more complex kettle of fish than what early-onset mutations served up. With early onset, all it took was the inheritance of one mutation, and the disease struck forcefully, like a sledgehammer. Its early strike and its clockwork-like descent in a family—these were blatant signs of the severity of its mechanism. "What you expect of an autosomal dominant pattern is that it won't skip generations and that, on average, 50 percent of siblings will be affected," describes Blacker. Hence, the disease's frightening regularity for every generation in its path.

Yet this type severity and recurring theme were noticeably missing in families harboring older-age cases. The genetic mechanism wasn't harsh enough to confer the disease any earlier in life. And by the same token, not every generation was affected. Moreover, in those that were, usually fewer than 50 percent of offspring got the disease—as far as one could tell. Assessing a late-onset disease's potential for recurrence in descending generations is problematic since older individuals who might have been genetically predisposed may have died of something else.

Still, the mechanism we were glimpsing in older patients wasn't consistent enough to be the handiwork of dominant mutations; nor, for that matter, recessive mutations. (Such mutations confer disease only when a person inherits the same aberrant copy of a gene from each parent.) The disease's late face was every bit as lethal as its younger face, yet it took so many years to strike that its fundamentals appeared more subtle than one dominant sledgehammer and more like several doctors' small reflex hammers hitting simultaneously—a muddle of multiple genetic risk factors impacted by aging, environmental, and socioeconomic factors.

An evolving picture of the DNA variations that lay scattered throughout the human genome gave us a better feel for our quarry. Scientists were guestimating that from person to person the human genome differs by approximately one-tenth of 1 percent. This may not sound like much, yet it amounts to a variation every 1,000 bases or so. Since the entire genome has roughly 3 billion bases, this amounts to between 2 and 3 *million* differences between any two individuals. As Eric Lander, a geneticist at MIT and Whitehead Institute, has rightly described it, "We are all the same, but we are all different."

One type variation—benign polymorphisms that have no bearing on disease—was now known to be plentiful. Presumably, most of the variations used for tracking diseases ever since the Huntington's project constituted this benign type. A second type—mutations that cause disease with 100 percent certainty—thankfully was considered to be a rare brute. And then there was the third type—a slimy hybrid, a polymorphism that lay somewhere between benign and causative and boosted a person's susceptibility for a disease. There was an increasing sense that this type not only commonly occurs in the population, but becomes all the more disease-conferring as a person ages.

Frequently I'm asked, What's going on? Why are so many people nowadays felled by heart disease, cancer, Alzheimer's, diabetes, and other illness? Apart from any environmental influences, it's apparent that as humans live longer, our longevity, like a receding tide, is exposing more and more of a family's inherent genetic bones—gene variants that predispose a person to age-related ills. Possibly if humans' average life expectancy were to lengthen by another two decades (since 1900 it has increased by nearly three decades), aging's lowering tide might expose never-before-seen genetic factors for disease—even hard-core mutations. Perhaps we would even find ourselves faced with new ailments and syndromes we haven't lived long enough to encounter.

As we geneticists waded further into the late-onset conundrum in the mid-1990s, other researchers were managing to gain slight clues about the environmental side of Alzheimer's late-onset equation. As will be seen, hints about certain external factors that can *detain* Alzheimer's were permitting a clearer picture of the disease's cascade of pathological events and, in turn, giving researchers a sense of which genes might be involved in late onset.

But first, what do we know about external hazards that can *accelerate* Alzheimer's dementia? One such hazard is head trauma—the third leading risk factor for Alzheimer's after aging and a family history. For instance, boxers are at risk for dementia pugilistica, a dementia brought on by repeated blows to the head. An unfortunate case in point is Sugar Ray Robinson, whose death from Alzheimer's surely was compounded—if not initiated—by his years in the ring. While head trauma can induce Alzheimer's cognitive decline, it can also produce a Parkinsonian-type motor deficiency, as experienced by Muhammad Ali. The differing conditions appear to depend on which regions of the brain receive the trauma, be it those important to memory or others central to motor skills.

Sam Sisodia and I recently had the honor of making the acquaintance of Muhammad Ali and his wife at the 125th running of the Kentucky Derby. As we sat at their table in the Sky Terrace waiting for the race to begin, I couldn't help but notice that although the Champ's hands shook and he had trouble speaking due to unruly larynx muscles, his presence of mind seemed very much intact. With great panache, he performed a magic trick, pouring salt from a salt shaker into the palm of his hand, then making it disappear. Impressively, the Champ still knew all the moves.

Interestingly, research indicates that a genetic susceptibility for a brain disorder can make someone all the more vulnerable to head-injury-associated neuronal loss. For instance, studies have shown that boxers with the APOE-4 gene variant are more prone to Alzheimer's-type dementia.

One last word about head trauma. Soccer being such a popular sport, there is currently much discussion about whether the practice of "heading"—sending the ball with one's head—can be mentally harmful. Recent studies don't exactly put one at ease. In tests that measure memory and planning abilities, soccer players were shown to underperform other type athletes. Observes Tracy McIntosh, director of the Head Injury Center at the University of Pennsylvania, "There's good lab evidence that mild brain injury, if repeated, can be very damaging. You see multiple pathological cascade features, and it's very scary."

To peer further into our environment, the suspicion lurks that toxins possibly accentuate genetic risks for Alzheimer's. Yet if certain ones do, they have yet to be proven guilty. This said, Ashley Bush, a biochemist in my lab, set forth a thought-provoking set of test-tube experiments in 1993 when he showed that the presence of zinc induces the clumping of A-beta into insoluble amyloid—dramatically so and in less than two minutes. He also detected that copper and to a lesser extent iron similarly increase A-beta's consolidation into amyloid. We couldn't help but wonder if these metals, therefore, were contributing to Alzheimer's. To use the zinc example, the amounts tested correspond to the micro amounts regularly seen in the brain. It wasn't necessarily that increased amounts of zinc from certain foods, medications, or other products were entering the brain. It simply might be that in older brains zinc, copper, and iron more easily go unregulated, resulting in their greater presence, which hastens A-beta's aggregation.

Today our evidence stands that zinc and copper could be strong promoters of amyloid formation, and therefore active disease participants. This observation is bolstered by the fact that others have shown that Alzheimer's amyloid plaque may be a flytrap for such metals and therefore possibly enlarge more rapidly because of them.

To mention yet another metal, time after time reports linking aluminum to Alzheimer's have sent gusts of fear through the public. Yet the current consensus in the field is that the aluminum threat may be more imagined than real. Sounding the alarm, past studies have pointed to higher amounts of aluminum in the autopsied brain tissue of Alzheimer victims, particularly in the plaques, than in brain tissue unscathed by the disease. These findings have made for loud headlines, but always in the background there have been a far, far greater number of tissue studies that have failed to find an aluminum excess in diseased brains or so much as a drop of convincing evidence that aluminum plays a role in Alzheimer's. It also should be noted that in Ashley Bush's abovementioned tests aluminum did not exert the same amyloid-inducing effect on A-beta as did zinc and copper.

Aluminum being such a ubiquitous substance in both its natural and processed form, the fear lingers in the public's mind that it might be seeping into the brain from environmental sources. Most researchers believe, however, that whatever small amounts a person ingests aren't easily absorbed by the gastrointestinal tract and get excreted, and that whatever traces do end up in the bloodstream probably are mostly barred from the brain by the blood-brain barrier—a system wherein cells in the brain's blood vessels are tightly packed, preventing too-large molecules from entering the brain.

As for foods that might augment a susceptibility to Alzheimer's, none has been formally charged, although one developing line of evidence is compelling. Recent studies have correlated high-cholesterol diets with an increased risk for Alzheimer's. The great harm of cholesterol in relation to dementia may, in fact, be its effects on accumulating A-beta in brain tissue, research suggests. When researchers at Eli Lilly fed transgenic mice a high-cholesterol diet, they observed a fivefold increase in amyloid in the brains of the mice. That bears repeating: a fivefold increase! If A-beta fibrils some day are proven beyond a doubt to be the toxic element in Alzheimer's, the current warning sounded by physicians may reach a new decibel: Beware fatty foods!

Indeed, in recent years accumulating studies provide evidence of a heart-brain connection in Alzheimer's. What's bad for the heart appears bad for the brain—which should surprise no one. "It's become fairly clear cut," maintains Mike Mullan at the University of South Florida, that "if a person is predisposed toward vascular problems, it can bring out Alzheimer's and that Alzheimer's in turn can bring out vascular irregularities." How the two are precisely related is undergoing rigorous examination. That the APOE-4 gene is associated with an increased risk for both atherosclerosis and Alzheimer's further supports a role for cholesterol in Alzheimer's. What we're left with, then, is that possibly past generations were not that far off in thinking that "hardening of the arteries" does hasten dementia, perhaps by permitting amyloid buildup. There's also reason to expect that a good cardiovascular flow has a detoxifying effect on the brain.

Abundant clues indicate that certain ethnic, cultural, and socioeconomic factors can influence Alzheimer's frequency. We need more research to sort out these factors, but one theory that holds more water than most is that the lack of a sustained formal education can lend to a greater risk. This has ushered in the "cerebral reserve hypothesis," which maintains that the more neuronal synapses a person develops, the more neuronal destruction has to occur before dementia occurs. The mental demands of education are thought to build up this synaptic reserve. Supporting this theory, researchers examining postmortem tissue found that older people who received more education had as much age-related brain shrinkage as people with less schooling, yet during their lifetimes remained more resistant to Alzheimer's than the less educated group. Not that a well-wired brain can stave off Alzheimer's for that long. All too often one hears of exceptionally erudite people—or people with steel-trap memories—who succumb to the disease.

Over the years, an endless number of other environmental suspects have been singled out as possible triggers of Alzheimer's dementia—the electromagnetic field emitted by power lines; mercury released from dental fillings; even too much consumed tofu—to name a few. But definitive evidence for these and many other claims is lacking. If convincingly identified, external hazards stand to be more easily controlled and mollified than inherited factors. The problem is, they tend to be extremely hard to pin down and aren't as immediately suspicious as, say, cigarette smoke is in relation to lung cancer. Scientists haven't been too terribly active about investigating environmental risks that might pertain to Alzheimer's, a challenge that deserves a much stronger commitment in the new century.

By mid-decade, science was making progress in recognizing the opposite situation—external factors that *detain* Alzheimer's, or at least appear to. Many of these are health and dietary supplements, and it's reasonable to expect that in the not-too-distant future some may emerge as proven placaters of Alzheimer's. For those of us tracking late-onset genes, what was so telling about several was how they mirrored various stages of the disease's cascade of events and offered hints as to which bad-acting genes might be behind the late-onset pathology. Let me say first off, each of these supplements requires more rigorous testing before their effects are entirely confirmed and understood. Someone contemplating their usage should consult a physician. Any health supplement or drug, even those sold over the counter, can have potential serious side effects.

This said, epidemiological studies have suggested that anti-inflammatory drugs such as ibuprofen, indomethacin (commonly used for arthritis), and other nonsteroidals may retard Alzheimer's onset by up to five or more years. Research in progress in John Breitner's Johns Hopkins lab shows that aspirin may be almost as effective. However, patient studies still are needed to provide concrete proof of the effectiveness of anti-inflammatories. Their possible advantage was first brought home when people taking medication for rheumatoid arthritis were observed to have a significantly lower rate of Alzheimer's. Separately, researchers were amazed to discover that patients in a leper colony in Japan who'd been receiving dapsone, which is both an antibiotic and an anti-inflammatory, similarly had a far lower incidence of dementia than those not on the drug.

This fit tongue-and-groove with what the field was learning about the stream of brain events that occur once something starts Alzheimer's ball rolling. Since the 1980s, researchers—notably Patrick and Edith McGeer at the University of British Columbia and Joseph Rogers, presently the president of Sun Health Research Institute—had observed that brains affected by Alzheimer's are grossly inflamed. Glial cells swarm to brain areas hit by the disease to try to clean up accumulating cellular and plaque debris, but once these housekeepers arrive on the scene, they actually activate the release of more nasty A-beta 42. This accelerates plaque formation, which attracts more microglia. And so it becomes a vicious circle. Most—but not everyone—agree that a robust inflammatory reaction seriously compounds the ongoing destruction of brain cells.

Studies mid-decade were also suggesting that vitamin E and other antioxidants can slow Alzheimer's progression. Antioxidants appear to have this positive effect because they help counter another phenomenon seen in the disease's molecular cascade—oxidative stress. In the deteriorating brain, excess amounts of the A-beta peptide as well as legions of microglia reacting to A-beta can boost the production of free radicals, oxygen-based molecules that mount a primitive defense. Yet their friendly fire results in the incidental kill of still more neurons, possibly leading to more free radicals, inflammation, and microglia. Antioxidants appear to partially assuage this chain-reaction situation with some success.

Apoptosis—the body's programmed death of cells—looked suspiciously like another confounding ingredient that exacerbates the cascade, although its role in Alzheimer's is highly controversial. Apoptosis—unlike necrosis, or uncontrolled cell death—normally helps the body rid itself of dysfunctional and unneeded cells. To some scientists it appears as if aggregating A-beta induces apoptosis, and that, in reverse, apoptosis induces the production of more A-beta. Mutations in the presenilin genes, it's been observed, may make cells significantly more prone to apoptosis, which possibly generates more A-beta and amyloid, which leads to more inflammation, which excites more free radical buildup, which produces more apoptosis—and so on and so on! Apoptosis remains a controversial call in neurodegenerative diseases because the cells it kills are rapidly cleared away by scavenging microglia, leaving no detectable bodies in the library. No trace evidence.

Yet another substance that appeared to detain Alzheimer's onset was estrogen used postmenopausally, or so epidemiological data suggested. Reports of women on estrogen have shown a 30 to 50 percent reduced risk for dementia as compared to women not on the hormone. Estrogen's ability to delay Alzheimer's shouldn't be banked on, however, until more conclusive data arrive. (A recent trial has shown, in fact, that estrogen may have no beneficial effect once Alzheimer's has reached a mild or moderate stage.) More women than men get Alzheimer's, and practitioners suspect it's because women lose estrogen and its positive health effects with menopause, while older men maintain fairly high levels of testosterone, some of which converts to estrogen. One recent study yielded evidence that testosterone actually might stymie Alzheimer's by lowering the generation of A-beta.

If estrogen is effective against Alzheimer's, what might matter is its demonstrated support of neurons and synapses. Also to consider are the many past studies that have shown estrogen's ability to elevate "good" HDL cholesterol, reduce "bad" LDL cholesterol, and generally benefit cardiac health, and therefore brain health. Although a recent large study has shown that estrogen taken for a few years had a very slight negative cardio effect on women, this runs counter to a multitude of longer-running studies that have reported the opposite.

The bottom line is pretty irrefutable: What is good for the heart is good for the brain. Anti-inflammatories, antioxidants, exercise, and even wine—each has been shown to benefit the cardiovascular system. And, guess what—all have been linked to a lower incidence of Alzheimer's. If there's an elixir for the brain, it appears to be the best natural cardio-therapy of all: exercise. In what comprises one of the most outstanding recent surprises in neuroscience, we now know that contrary to what was once thought, an adult brain can make new neurons. As observed in mice, new neurons in the memory-enhancing hippocampus especially bloom when mice take to the running wheel. Could it be that someday the "reserve hypothesis," which correlates a lower incidence of Alzheimer's with more years of education, will be extended to those who have exercised more throughout their lifetimes, thereby building up their neurons and synapses?

Accounts of less investigated therapies for Alzheimer's constantly pop up. These include medicinal plants such as cat's claw, gotu kola, Siberian ginseng; extract from the ginkgo tree; scalp acupuncture; folic acid from dark-green leafy vegetables; blue-green algae; black moss; homeopathic mixtures and other ancient prescriptions. That large drugmakers are paying closer heed these days to the local lore surrounding various botanicals surely is one of the most promising developments in drugmaking.

The more researchers grasped about supplements that can slow Alzheimer's, the more this knowledge filled in a picture of how Alzheimer's pathology progresses once "the marble rolls off the table," as one colleague describes it. Nudging the marble off the table in the first place are, in early-onset cases, lone-acting mutations; or in late-onset cases, most likely a mixture of risks tied to aging, genes, and the environment. Both scenarios appear to provoke a troubling accumulation of A-beta fibrils, which the amyloid camp views as the primary destructive agent in Alzheimer's—the toxic killer of neurons. Others, on the other hand, perceive accumulating A-beta as a marker for some altogether different lethal insult. Once the cascade is launched, A-beta, amyloid, and cell death trigger an explosion of other phenomena. Many of us consider the tangle a later-stage event, the signature of a dying neuron, especially as a number of studies show A-beta congregating well before tangle formation.

This stream of events, the essentials of which appear below, shouldn't be construed as a linear progression that's set in stone. Just about every lab has a different interpretation of what gives way to what.

Pathological Event		Potential Therapies
Gene mutations (early-onset); multiple genetic, aging, and environmental factors (late-onset)	→	Detection screening to identify people at risk and those who require early treatment
Processing of APP protein generates A-beta 42	→	Compound to prevent APP's conversion to A-beta 42
Accumulation and aggregation of A-beta 42	→	Compound to inhibit A-beta aggregation
Inflammatory response	→	Anti-inflammatory drugs
Oxidative stress	→	Antioxidants
Apoptosis	→	Specific protein inhibitors
Brain chemical deficiencies	→	Cholinesterase drugs (e.g. Aricept)
Neuronal degeneration and tangle formation	→	Neuroprotectants, (e.g. estrogen) or transplantation of neuronal cells

FIGURE 11.1 Likely cascade of pathological events in Alzheimer's.
Illustration: Robert D. Moir

It was tremendously encouraging that potential subduers of Alzheimer's pathology—like anti-inflammatories or antioxidants—already lay within society's medicine chest. The modest goal of delaying the disease by five to ten years seemed that much more obtainable. As one journalist observed, if the disease's entrance later in life could be temporarily held in check, "that would allow us to die quietly and nicely from some other disease." Slowing Alzheimer's by just five years might reduce the number of cases by 50 percent; ten years—by a remarkable 75 percent.

Therapies aimed downstream in the cascade conceivably could go a long way toward achieving this goal. But drugs aimed high upstream—directly at A-beta, either as it was being clipped from its bigger protein or clumping into amyloid—might prove the most effective route of all. Envisioned were drugs that could lower the production of A-beta just as treatments for heart disease keep down bad cholesterol.

All sorts of features in the disease's unremitting cascade were seen as grounds for suspecting certain late-onset susceptibility genes. Under surveillance, for instance, were genes connected to inflammation, oxidative stress, neurotransmission, and cholesterol metabolism—the latter possibly lending to amyloid's production. Those of us who believed that the A-beta fragment was the disease's destructive element expected that an important litmus test for late-onset genes would be that, similar to early-onset genes, they too would lead to radically higher levels of A-beta in the brain.

By this time, the idea that APOE-4, the one captured late-onset gene, indirectly caused the disease by allowing tangle formation had gone stale. In fact, there were solid signs that APOE-4 promoted A-beta aggregation. Every Alzheimer's gene so far identified, therefore, appeared to be a party to amyloid buildup.

At Mass General, Marilyn Albert's and my labs were putting our trust in classical genetic-linkage analysis. The same tool had been invaluable in handcuffing the three early-onset genes. It should work again for late-acting genes. The only drawback was that the NIMH-sponsored Alzheimer's initiative had collected very few large late-onset families with multiple affected members. Alzheimer's late variety was just too erratic to affect that many in a family. The number of small families collected was vast nonetheless, which compensated. The majority included two to five siblings and/or first cousins known to have come down with the disease after the age of sixty.

Only a handful of other teams, including the Duke team, Jerry Schellenberg's Seattle lab, and a collaboration between John Hardy at the Mayo in Jacksonville and Alison Goate at Washington University, similarly were employing real genetics for tracking late-onset genes. Elsewhere, dozens of university labs were using an entirely different scheme that utilized case-control studies. Bypassing classical genetic-linkage methods, many investigators were selecting a favorite candidate gene on the basis that its protein conceivably might contribute to Alzheimer's inflammation, oxidative stress, or some other event in the disease's downhill cascade. They then focused on a variant of that gene known to commonly appear in the population at large and tested it in Alzheimer patients (cases) versus similarly aged healthy individuals (controls). If a gene variant was, say, 1.5 times more present in patients than in controls, lickety-split they wrote up a report on their newly discovered Alzheimer gene.

By mid-1996, case-control studies had spawned proposals for close to a dozen late-onset suspects. Yet in most instances follow-up studies punctured the data. Very possibly in the future some of these candidate genes will pan out; the ACT gene on chromosome 14, for instance, is currently under strong suspicion. But in my view, case-control tests often draw on too few cases and controls to yield reliable results. If it's true that multiple genes provoke late-onset Alzheimer's, a group of a hundred randomly selected people may be too random, too genetically heterogeneous, and too small to divulge a true statistical measurement of a gene's risk potential. Thousands of people from many independent studies need to be tested to rule out spurious results.

Sticking to our more classical genetic-linkage strategy for unearthing genes, in the spring of '96 our MGH group began analyzing the large cache of family DNAs we had access to. Toward summer, an old feeling of luck in my bones began returning. Not just one but two of the chromosomes being tested—3 and 12—were signaling risk genes. DNA markers were cosegregating with people who had the disease! The Fifth International Conference on Alzheimer's Disease and Related Disorders would soon convene in Osaka, Japan, and we pushed hard to flush out more data in time to present the news.

It was a close shave. It wasn't until two hours before my scheduled talk in Osaka that Marilyn Albert, phoning from Mass General, reached me in my room at the Royal Hotel and shared the final numbers. Jet-lagged and pumped by coffee and the excitement of announcing our findings, I was decidedly distracted by the time I reached the lecture hall and became all the more so when an attendant knocked the water glass off the lectern and began sweeping up the shards so assiduously that his fingers began bleeding. Being one to look for omens, I saw this as a distinct rain cloud. Before several hundred listeners, I announced our hits on chromosomes 3 and 12, whereupon the rain cloud opened. Peggy Pericak-Vance stood up to note that her Duke University team, by way of genetic linkage analysis, also had hold of late-onset evidence on chromosome 12.

There was that familiar sound of hoofbeats again. Why is it that whenever something interesting turns up—*clomp, clomp, clomp*—contenders turned up as well? For the sake of trapping the truth and making progress, this clomp, clomp, clomping only could be regarded as a good thing, however. Very certainly it was speeding us forward.

The hoofbeats in our own genetics corner, in actuality, were tame prancings compared to the thunderous clompings that had begun noticeably shaking the pharmaceutical universe. The riders included everyone from huge swashbuckling drug companies to the smallest biotechs. Spurred on by their various visions for an effective Alzheimer drug, they were galloping forward for all they were worth. It had all come together for this global phalanx in the last few years—the insights into the biology, the strong incentives, the technical inventions. For many, the most attracting force of all was the amyloid hypothesis.

Within a year of Fran's first noticeable symptoms, another sister began showing signs of chronic memory loss. Having watched their mother's incapacities emerge, the Noonans recognized the signs—"the changes in personality, the focusing on little hurts and not getting beyond them, the paranoia, the forgetfulness, and that blank look"—in the words of an older sister, Pat. Because of their mother, their "antennae" were always up. A grieving was setting in that would go "on and on. We would find ourselves continually revisiting our losses," notes Pat. Many would seek counseling for the depression they felt. They were realizing that since there was no escape, they wanted to mount an offensive. It might have been different if theirs was the only generation facing Alzheimer's. But between them, they had twenty-two children and one grandchild, with numerous other grandchildren on the way, all of whom would be just as vulnerable. It was due to their children's fate, and the children to follow, that they decided to contribute their DNA to researchers at Massachusetts General Hospital. Their blood samples might help set their family free.

— *twelve* —

A Gamble for Hope

Nothing is secret, that shall not be made manifest; neither any thing hid, that shall not be known and come abroad.
—Luke 8:17

Even as DNA's spiral was being scanned for more late-onset Alzheimer genes, a fervent chant had sprung up across the research community. *Stop A-beta and you'll stop Alzheimer's disease*. Not everyone believed this drug approach would work, but enough scientists did that it was creating amazing activity in the pharmaceutical and biotech industries. Treatments aimed at oxidative or inflammatory stress might diminish the disease's downstream torrent in the brain, but to nip the pathology in the bud, one wanted to deliver a blow as far upstream as possible, and A-beta—and its accretion into amyloid—was the most upstream target in sight.

Similar to any drug, an A-beta inhibitor would require a wing and a prayer, not to mention a minimum investment of $200 million and a concept-to-market time of at least ten years, if lucky. An estimated 90 percent of conceived drugs fail, never reaching the end of a company's long pipeline. An A-beta inhibitor—sprung from a hypothesis, a supposition, a still-tentative assumption—fell into the riskiest-type launching. A-beta might look guilty, since each discovered gene aberration led to its accumulation and aggregation in the brain—brains wherein neurons die. But then again, we still had no final proof that A-beta was the disease's primary killer, its Grim Reaper. No better theory existed, however, at least none that had universal acceptance. And meanwhile the drug industry was under terrific pressure to do something about a disease that afflicted an estimated 12 to 14 million people worldwide. Very possibly this figure was a good deal higher since so many cases were known to go unreported.

The early months of 1997 found SIBIA Neurosciences—a small biotechnology company situated a stone's throw from the Pacific Ocean and within an easy stroll of La Jolla's shop-filled center—on track toward an A-beta inhibitor, albeit in a very preliminary way. SIBIA and one of its pharmaceutical partners, Bristol-Myers Squibb, had closed in on an interesting "lead," a rookie compound that showed promise in a test tube but faced tremendous hurdles if it was to achieve its desired effect in the brain. After undergoing rigorous tinkering, it first had to prove itself in preclinical tests, in both the test tube and mouse. Then, if it was to receive FDA approval, it had to successfully complete three phases of clinical trials in humans.

To go the distance, SIBIA's president and CEO William Comer was all too aware, would take impeccable precision, a long patience, and more than a thimble of luck. It wasn't as if their compound simply had to be gotten to the gut—they had to get it all the way to the brain's intricate interior, and the closer to that remote hemisphere they got, the more variables their compound would be up against.

A chemist and pharmacologist who before coming to SIBIA had worked at Bristol-Myers Squibb for some thirty years, much of that time spent directing Bristol's worldwide research-and-development sector, Comer had seen more than a few compounds scuttled as far out in the pipeline as phase two of human clinical trials, the stage at which testing of usually a few hundred individuals puts a drug's efficacy and safety on the line. On the other hand, he'd directed the full passage of notable successes, drugs such as BuSpar, Bristol's antianxiety medication, and Taxol, presently America's number-one selling cancer drug. In other words, Comer wasn't your average biotech CEO; he wasn't just interested in the dollars-and-cents side of the equation. "Bill was unusual in that he was fascinated in—really passionate about—the science," says Maria Kounnas, a biochemist who was directing SIBIA's test-tube studies of the A-beta compound.

I was aware of what Comer and SIBIA were up to because I'd kept in close contact with Steve Wagner—the director of SIBIA's Alzheimer program—ever since we'd met in Maratea, Italy, nearly a decade earlier. Wagner, Sam Sisodia, and I had nicknamed ourselves "The Triumvirate." Our skills aptly complementing one another's, we'd optimistically decided one night over too many beers that between the three of us we should be able to solve Alzheimer's. Being the geneticist, I'd try to further come to grips with the *where* and *what*—where and what were other genes that contributed to the disease? Sisodia, the cell biologist, would figure out the procedural *how*—how, exactly, does the pathology proceed? Waggie, the biochemist, would supply the ultimate *how*—how can the pathology be halted? (Even though this book is full of *whys*, in formal biology there are no *whys*, for you can't ask why a molecule does such and such, since as intentional as a molecule may seem, it's void of conscious intent.)

Our rambling discussions along these lines continue into the present, spilling out in restaurants and ale houses wherever in the world a meeting brings us together. All too often we think we've got the problem licked—what causes A-beta's buildup and how to stop it—only for new data to emerge, necessitating a total revision. "We have a good time, but we're also ready to challenge each other. If we get to disagreeing too much, we just have another beer," describes Sisodia, now chairman of the Department of Neurobiology, Pharmacology, and Physiology at Johns Hopkins.

Needless to say, the field still had infinite miles to go before one question stopped leading to another. If they ever did. Nevertheless, the Bristol-SIBIA collaborators, like numerous other drug designers, had decided to spring off the existing evidence. They had chosen as their drug target the gamma-secretase, one of the two unidentified enzymes thought to release A-beta. Slow down gamma, and you might delay A-beta's production. Hence they were working from a twofold assumption: that excess A-beta was Alzheimer's deadly protagonist and slayer of neurons; and that gamma-secretase, which existed in name only since no one had isolated it yet, was instrumental in setting A-beta loose.

The vision for this particular plot against Alzheimer's had emerged in late 1990 during a blinding snowstorm in Louisville, Kentucky, three months before Comer and Wagner arrived at SIBIA. Wagner, then at UC, Irvine, had been involved in a start-up company geared toward developing a diagnostic to detect A-beta's release, and he was on the lookout for someone to help guide and fund the research. Comer, then at

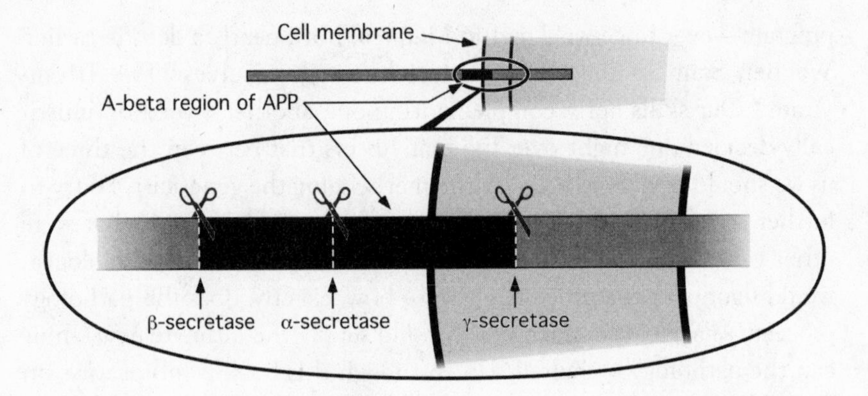

FIGURE 12.1 APP protein and secretase sites. Bristol-SIBIA collaborators took aim at the gamma-(γ) secretase. Illustration: Robert D. Moir

Bristol, was thinking of retiring from Bristol for the sole purpose of throwing himself into a company dedicated to finding therapies for Alzheimer's. He was disappointed with the drug industry's slow pace in regard to treatments for neurodegenerative diseases. Accentuating his outlook, he was witness to his mother's late-stage Alzheimer's and her worsening vegetative state. His father wasn't any better off. He was blind, afflicted by a Parkinson's-like muscular disorder, and confined to a wheelchair due to a faulty hip replacement. Never before had Comer, who had been "born wanting to discover drugs," gotten such a troubling view of medicine's limited pharmacopoeia for people with age-related disabilities. How euphemistic to think that people merely die of old age, for plainly his parents were failing from conditions separate from yet compounded by aging.

A mutual acquaintance put the two men in touch with each other, and over Christmas vacation, Wagner having returned to his home state for the holiday and Comer being across the border in Evansville visiting his in-laws, they took the opportunity to meet in the lobby of Louisville's Hyatt Regency. As the blizzard outside whirled white, inside they talked shop for four straight hours. "At our first meeting, the idea of going after diagnostics and therapeutics for Alzheimer's just crystallized," recounts Wagner. "Bill had infinite experience in how to get a drug all the way through the FDA and in team-making. I had expertise in APP and A-beta."

Recalls Comer, "Drug companies were making fourth- and fifth-generation antibiotics, or third- and fourth-generation antidepressants, and they simply weren't going out on a limb when they didn't know the cause of a disease. They especially were staying away from neurodegenerative diseases, because they didn't know where to begin and it took too much of a financial risk. Yet the academic medical schools and labs were providing clues, and though nothing was proven, Steve and I were in agreement that it was worthwhile taking that thinking into a drug discovery mode—to go after Alzheimer's by slowing down the production of A-beta. Better to have compounds than hypotheses."

Comer, a fit twenty years older than Wagner, can be every bit as much an athlete in the boardroom as he is on the tennis court. When talking about drug invention, he exerts an unbridled optimism, as does Wagner, and as do most of us. The problem-solving associated with drugmaking has a way of eliciting a no-holds-barred "this-can-work" attitude, and one can just imagine the contagious energy the two men generated that snowy afternoon.

After Comer and Wagner's initial meeting in late 1990, a great deal fell into place in a very short time. SIBIA's directors were looking for new leadership and direction for their company. One of the country's oldest biotech establishments, SIBIA had been founded in 1981 by Salk Institute president Frederic de Hoffmann—thus the acronym SIBIA, for Salk Institute Biotechnology/Industrial Associates—and this "first child of Salk," as Comer calls it, had been doing contractual research for big companies, never really blossoming into its own programs. Its board, looking to rejuvenate SIBIA, ended up not only purchasing Wagner's fledgling company and establishing Wagner as a scientific director, but also installed Comer, steeped in R&D experience, as SIBIA's new president and CEO.

Thus a new SIBIA Neurosciences was born, one that would develop primarily small-molecule drugs that could surmount in-body barriers more successfully than big-molecule drugs. Approximately 50 percent of its drug-probing muscle would go into Alzheimer's, with additional projects pitted against an array of other neurodegenerative conditions. Critical to its progress were state-of-the-art bioassays and automated methods. *High-throughput screening*—the rapid testing of large numbers of compounds to evaluate their ability to modify a specific biological target. *Combinatorial* chemistry—the rapid synthesis of novel compounds. *Medicinal* chemistry—the tailoring of a compound's structure to ensure its optimal safety and distribution in the body and its optimal selectivity and potency when hurled at a specific biological function. These capabilities are indispensable to today's rapid-fire world of drug discovery. SIBIA, an older biotech, would notch up its strength in each area.

Then along came the 1992 landmark discovery that persuaded so many drug establishments to get serious about pursuing A-beta. As suddenly realized, the A-beta peptide wasn't just an abnormal brain by-product derived from a cellular goof, but a molecule normally churned out by cells throughout the body. Only when it accumulated in the brain could it be considered a menace. Before, presumed to be a brain-only disease-related phenomenon, it had appeared hard to access. But now researchers could take practically any cells from the body, grow them in a dish, toss compounds at them, and hope to find at least one compound that stymied the cells' release of A-beta.

To Bristol-Myers Squibb and other big pharmaceutical companies, this was an all-important green light. "It changed our focus," notes Kevin Felsenstein, a molecular biologist and principal scientist at Bristol. "We had only about four biologists working on Alzheimer's, but once we realized we could measure A-beta, this said to us, We should try to design a screening paradigm" for an A-beta inhibitor. (At some big pharmaceuticals, upper management deemed the pursuit of A-beta far too risky and held chemists back from trying.) Additional chemists were recruited, and Bristol's high-throughput team, housed at the company's Wallingford, Connecticut, research facility, began randomly screening Bristol's library of hundreds of thousands of compounds for a molecule that might inhibit the action of either the beta- or gamma-secretase that presumably freed the A-beta fragment. Similar to other big houses, Bristol had collected a wealth of chemical compounds from myriad projects over the years, most synthetically spun in the laboratory rather than naturally made. Before long, Bristol's chemists were scoring "hits"—potentially interesting compounds that turned down A-beta's production. Those found appeared to primarily deactivate the gamma-secretase's cleaving of APP.

Across the country, SIBIA similarly had started riffling through its more modest library, primarily for molecules that might impede the *beta*-secretase. This is the same secretase George Glenner had imagined he had by the tail, although after his death his lab never came up with conclusive proof. Shopping around, as biotechs do, for a big, solvent drug company to back their work, SIBIA impressed Bristol. "The feeling was, SIBIA was a good mesh for us," says Felsenstein. "They were working on the beta-secretase almost exclusively, whereas our compounds had led us to gamma. So here were two parallel programs devoted to stopping A-beta's processing. A collaboration could give Bristol more manpower than we had internally and help accelerate our A-beta project."

Consequently, big Bristol formed a corporate partnership with small SIBIA in the summer of 1995. Kevin Felsenstein would supervise the operative, reporting to Perry Molinoff, the vice president of Bristol's neuroscience drug discovery sector.

Bristol was hardly alone. By mid-decade, Novartis, Eli Lilly, Pfizer, Parke-Davis, Pharmacia & UpJohn, SmithKline Beecham, Hoechst Marion Roussel, and most every other pharmaceutical giant were at one stage or another of pursuing A-beta and/or amyloid. Like Bristol, some had chosen a biotech partner to facilitate the research. The attempt to interfere with the secretases' clipping activity was only one strategy being employed. Other schemes consisted of finding compounds that might prevent already-excised A-beta from aggregating into amyloid, or enhance its degradation, or promote its clearance from the brain.

Numerous houses were spreading their bets by developing, along with one or more A-beta drugs, compounds directed at totally different Alzheimer-related targets. Drugs to preserve or replace brain chemicals. Drugs to enhance neuronal growth factors. Drugs to subdue brain inflammation. Drugs to control calcium streaming into disease-hit neurons. Drugs to quell Alzheimer's free-radical cerebral rampage. On the way as well were strategies to replace lost brain cells with neural stem cells, nascent cells that become neurons. The more targets and avenues sought, the better a company's chances of at least a partial bull's-eye. But many of these efforts, if successful, would simply reduce symptoms and the downstream sequelae of pathology. The A-beta approach was the true pièce de résistance, since, as perceived, A-beta was the upstream solicitor of so many downstream woes.

Several young biotech companies were so enticed by A-beta that they had scrounged up considerable wads of venture capital and were going it alone, without a corporate partner. In most instances, the hoped-for payoff down the line was that if their screening program or emerging drug showed merit, a major drug company would buy them up and whisk their operation off in a market-bound sparkling-gold chariot. SIBIA, so differently, had never raised a nickel of venture capital. Instead, it sought capital through contracts with large pharma, its directors opting for a smaller percent of profits rather than dilution of ownership.

The eagerness on the part of so many businesses, big and small, to bet high stakes on curbing A-beta isn't so very mystifying. A rule of thumb in today's drug industry dictates that if a speculated product looks as if it won't achieve $100 million in gross annual sales by its third year on the market, it's not worth developing. Expectations for a useful Alzheimer drug well surpassed this. "It's unlikely that an effective Alzheimer's drug would be a billion-dollar drug right out of the chute, but if its efficacy, safety, ease-of-use, and other criteria hold up, its annual peak sales could eventually be in the range of greater than 1 billion," says Larry Altstiel, a former Eli Lilly senior scientist. Others forecasted even higher gross annual sales—in the range of $2 to $6 billion. Among all the drugs available in the world, only a handful achieve sales of over $1 billion. Prozac, for instance, the antidepressant made by Eli Lilly and one of the drug industry's benchmark success stories, brings yearly returns in the neighborhood of $3 billion.

The drug industry almost couldn't afford not to gamble on A-beta. "That big pharma needs to improve the flow of new drugs is indisputable," the *Economist* would report in 1998. In order "to maintain their current annual revenue-growth rate of 10 percent without resorting to yet more mergers," the top ten pharmas "will have to increase their productivity tenfold." No small accomplishment. Yet what if, in the case of an A-beta inhibitor, the prodigious efforts and billions of dollars being aggressively applied to a mere hypothesis, a supposition, a tentative assumption, went kerplunk? What if A-beta blockers didn't achieve their intended goal?

"Ten years down the line, we could look back and say, boy, were we stupid," remarks Leon Thal, chairman of neurosciences at UCSD. "Alzheimer's pathology *appears* to all link back to the misprocessing of A-beta. Nevertheless we don't know if Alzheimer's happens due to one single cause or many." Aging plus ethnicity plus environmental hazards plus A-beta buildup plus abnormal tau plus inflammation plus oxidative stress—maybe the whole ball of wax creates the pathology.

Despite that, most of us saw A-beta as the chief toxic malefactor. The mark of a strong hypothesis is that it's ready to be put to the test, and when Bristol and SIBIA joined forces in mid-1995, the amyloid hypothesis felt very ripe in that respect. Although we couldn't be completely positive that the A-beta peptide was the disease's toxic element, a popular theory proposed that just as these peptides start to self-aggregate into small fibrils, they are small enough to access the nooks and crannies of a neuron's outer membrane and thereby initiate a death mechanism inside the cell. As fibrils aggregate further, they might become too big to gain such a lethal foothold and lose their toxicity. Groups had sprung up that spent all their time studying A-beta aggregates, with special attention paid to the longer, seemingly more deadly A-beta 42 that clumped so rapidly.

Among amyloid researchers, there was debate about whether the plaques themselves were toxic. Many—myself included—felt that once A-beta formed mature plaques, A-beta lost toxicity. But others believed that plaques represented a potentially lethal pool of toxicity.

Pushing the amyloid hypothesis into testing, enough of the biology was there, and more than enough financial incentive. But perhaps most hard-felt was the rapidly swelling medical imperative to find an effective course against Alzheimer's. "The FDA had been under pressure for some time," observes Bill Comer. "Alzheimer's had gone to the top of the list of societal ills." Accounts of well-known people affected by Alzheimer's had continued to serve as a sobering advertisement of the disease's fierceness. When on November 5, 1994, former president Ronald Reagan delivered his handwritten letter to the public stating that he had Alzheimer's, it had brought recognition of the fatal disease to new heights.

Because older Americans represented the nation's fasting growing segment, Alzheimer's was continuously on the rise. Some speculated that it might soon move from the fourth leading cause of death into third place, overtaking stroke, since medications that thwarted stroke were proving so effective. Unless an effectual treatment was found, the number of Americans with Alzheimer's would triple, it was estimated, to a staggering 14 million by 2050 or sooner. Some projected that

Alzheimer's would become the most rapidly accelerating disease in the developed world. Baby boomers were beginning to turn fifty, and ahead in their path they saw Alzheimer's as their dreaded Ebola.

(As currently reported, 4 million Americans have Alzheimer's, compared to a reported 12 to 14 million worldwide. The contrasting low worldwide figure may reflect the underreporting of cases in other countries, and/or the fact that the U.S. has an older population than many countries, according to the Alzheimer's Association.)

It went without saying that Alzheimer's emotional cost to patients, their families, and other caretakers was immeasurable. What could be measured was its related healthcare costs, which for Americans fell in the vicinity of $100 billion per year. This price tag, it was being noted, obviously would soar as the disease's prevalence soared. Many times greater were the indirect dollar costs—a business's loss of productivity, for instance, due to the time employees take off to care for an afflicted family member.

In short, every day it was becoming more apparent how extensively the disease was bankrupting families, straining nursing homes, bleeding the nation's workforce, and making a mockery of a healthcare system wherein neither federal, nor state, nor private providers began to cover expenses for the round-the-clock, multiyear care Alzheimer patients require. "The availability of funds for Alzheimer patients is very low. Very few qualify for Medicare or Medicaid," today notes Paul Raia of the Massachusetts Alzheimer's Association. That over 70 percent of patients are cared for at home, usually by family members and for many, many years, presents a very dismal, haunting picture of people who supposedly have reached their "golden years." Observes Raia, "In a typical at-home situation, a seventy-five-year-old woman is caring for her husband—lifting him into the shower, feeding him, changing his diapers, and all the while grieving the loss of the person she once knew."

On the drug scene, the only Alzheimer drugs so far approved by the FDA—cholinesterase inhibitors—left a lot to be desired. Twenty years had passed since researchers had retrieved the first major clue from the disease's biochemical slurry—that levels of acetylcholine drastically decline in the brain due to destroyed cholinergic neurons. Emerging from the gruelingly slow pipeline, the first cholinesterase inhibitor, Cognex, had reached the market in 1993. A second, Aricept, made by Eisai, a

Japanese pharmaceutical company, and licensed to Pfizer in this country, had just received FDA approval in late 1996. It had fewer toxic side effects than Cognex and worked better for some patients. But by now drugs that treated a neurochemical deficiency took a back seat to those that might prevent neuronal loss. Even so, nearly a dozen other cholinergic drugs were slouching through development, some about to reach daylight. Second-generation versions were anticipated to supply a more potent wallop of acetylcholine to brain regions most severely affected by the disease.

In addition to its Bristol-backed A-beta inhibitor, SIBIA was working on an acetylcholine booster. Based on its performance in rats, it was expected to deliver a twenty-five-fold increase of acetylcholine in the hippocampus in comparison to Aricept's far lower impact, according to Bill Comer. The hope? That it would help a patient make the best use of what they have left. "If a patient has the disease, it may be too late to treat the underlying pathology," says Comer. "With an acetylcholine drug, we can at least fire up remaining neurons and try to increase the connections between neurons and maybe salvage some."

On sped the screening of Bristol's jumbo library. On sped the search for a compound that could block the gamma-secretase's activity without researchers knowing its identity, in order to subdue a toxic perpetrator—A-beta—without clear proof of its toxicity! Bristol's high-throughput team had the capacity to screen thousands of compounds in a week. (By the decade's end, like others of its ilk, it would be able to screen as many as 100,000 per day.) How different from just a decade earlier when a chemist, beholden to the manual mixing of chemicals, was lucky to test the viability of thirty or forty compounds within a year—often further slowed down by first having to laboriously synthesize the compounds.

It wasn't long before Bristol's chemistry section, headed by David Smith, fastened onto several new "hits" for gamma-secretase—potentially desirable compounds. "Normally for every 10,000 compounds tested, you expect around one hit," explains chemist Ben Munoz, who was serving as Smith's counterpart at SIBIA. "You don't get many, but the hits you do get are usually good ones, and if you're part of a first-rate group of scientists, you're off and running very quickly."

Two of the hits seemed extemely promising. Keeping one in-house to advance, Bristol gave the other to SIBIA for further development. There, under the direction of Steve Wagner, director of biochemistry, and with frequent communication between the big and small companies, SIBIA's chemistry, in-vitro (test tube), and in-vivo (cells and mouse) teams swung into action. Turning a "lead" hit into an ingested drug that reached the brain, deactivated the gamma-secretase, and lasted many hours would require endless fiddling on the part of medicinal chemists. Assisted by "Homer" and "Lisa," two computerized workstations, more than likely they'd have to revise their compound scores of times—a never-ending process of substituting and appending chemical units—if it were to have the desired effect in cell cultures, then mice, and finally humans. "You'd like to think all of this is done out of brilliance, but it's not," notes Maria Kounnas. "It takes a lot of hard work—and a little luck. Without a little luck, you can work yourself to death and never get there."

"The challenge is to shape a molecule so that, like a key, you can jiggle it through several different tumblers," observes Munoz. "You don't want to pick a lock. You want to have a key that slides in all the way and opens one door, not several." The body's most formidable hurdles include the liver, the body's chief detoxifying organ, which can render a drug ineffective or be damaged by its toxicity; and the blood-brain barrier that so effectively stops oversized molecules from entering the brain. "The brain-blood barrier is unforgiving," notes Munoz. "It protects the brain, but it becomes a liability for drug design. You can make fantastic inhibitors, but they aren't necessarily going to cross that junction." Add too many side chains to a compound to assist it past other obstacles, and it might run aground. Even if it got into the brain, it still had to survive dispersal and reach neurons in the hippocampus, cortex, and other regions attacked by Alzheimer's. Then one had to hope it escaped certain organelles in cells that degrade molecules, or else—"bang, your drug is gone," observes Munoz. And, finally, once within striking range of gamma's mechanism, it still had to have enough selectivity and potency to impede its target's action.

"No one had ever done anything like this before—tried to deliver a drug all the way from the mouth to the brain that could make a protein fragment go away," notes Steve Wagner. Typically, drugs meant for the brain were intended to either lower or increase the levels of various brain chemicals.

In the early months of 1997, the latest in a series of compounds being worked on at Bristol and SIBIA were demonstrating the proper activity in the test tube. Cells weren't making half as much A-beta as before. Holding their breath, the in-vivo teams at both establishments started testing these latest versions on mice engineered to overproduce A-beta, their brains spotted with amyloidotic spheres.

In early 1997, researchers at Athena Neurosciences in south San Francisco similarly were laboring over compounds they hoped would arrest Alzheimer's A-beta peptide. They'd let it be known that they had drug programs under way devoted to both the beta- and gamma-secretases. In fact their gamma quest, in conjunction with Eli Lilly, had been going on for several years, although both companies remained very hush-hush about it. Purchased in '96 by Elan Corporation, which was headquartered in Dublin, for the time being Athena represented a free-standing division of Elan.

When Kevin Kinsella founded Athena ten years earlier, he'd done so with the express purpose of developing Alzheimer-related diagnostics and therapies. Even the name Kinsella had chosen for his company—"Athena," the goddess sprung fully formed from the head of Zeus—was in pointed reference to the brain. Its early leap of faith and its belief that Alzheimer's was treatable had gained Athena infinite respect during its relatively short history. I'd served as a consultant in its early days and was fascinated, as were so many of us, by what Athena might achieve. During this tenure began what would prove a long-lasting friendship with one of its senior scientists, Dale Schenk. I quickly realized that if Schenk's strategic skills as a biochemist came anywhere close to his tournament-level mastery of chess, the disease had better watch out. Our very first game, both of us bet everything we had—everyone in our labs, all of our lab equipment plus personals, right down to our shoes—and he annihilated me in five minutes flat. We've kept up play ever since—via E-mail or cramming in games on a portable board at conferences. Frustratingly, I've never beaten Schenk, just stalemated him—and only once.

Athena's efforts, by 1997, had broadened to other diseases, but in the meantime its scientists had contributed several firsts to Alzheimer research. Among them, the finding that A-beta was a normal, not an errant, peptide; and the theory that a secretase must beneficially snip A-beta in half, which suggested the presence of two other secretases that liberate intact A-beta. One sensed that progress was inevitable, especially since Athena had in its keeping the field's first amyloid-generating mouse model, a proving ground for any drugs it might be concocting.

Of late Athena again had been in the news because of its launching of the first commercial diagnostic test for Alzheimer's. It was based on APOE-4, the gene variant that significantly raised a person's risk for late-onset Alzheimer's. Athena had licensed Allen Roses and colleagues' APOE patent from Duke University. APOE-4 testing had been carried out on patients in the research setting for a few years, but now that it was commercially available, it was feared the test could be misused since APOE-4's presence in someone's genome isn't a definitive marker for Alzheimer's. In the spring of '96, a *Lancet* report by an NIA and Alzheimer's Association–sponsored working group, for which I served as one of the committee chairpersons, laid down guidelines. It concluded that the test be given only to those with Alzheimer-like symptoms in an attempt to differentiate what type of dementia a person had, and furthermore that the test be given in combination with proper genetic and psychological counseling. By no means was APOE-4 to be used as a sole diagnostic for the disease.

Even when the APOE-4 test is used to differentiate forms of dementia, the rationale behind it remains weak, I feel. If someone has this gene variant, it might be fuller evidence that his or her dementia is due to Alzheimer's, but it's no guarantee. With nothing solved, a person and his or her family can wind up all the more anxious. If APOE-4 is in a person's bloodline, an added worry could be that other family members might be at risk—even though, in reality, they might not be.

In addition to its advancing secretase drugs, in the early months of 1997 Athena was just embarking on a far different—some would have said whoppingly speculative—approach to contain A-beta in the brain. The plan had originated quite a ways back when a half-dozen or so of Athena's senior scientists, including Dale Schenk, had gotten together for a think-tank session. "We were trying to wrack our brains about other therapeutic approaches we should try," recounts Schenk. "I was thinking to myself, I wish we had a way to put a sponge in the brain to tie up amyloid. A vaccine would be great. Too bad that the blood-brain barrier wouldn't allow it into the brain." Because the dogma said as much, no one had ever attempted a vaccine for a brain disorder, according to Schenk.

He dared share his whim with his colleagues. What if they tried immunizing their transgenic mice against amyloid? They could inject human A-beta 42 into their hind muscle, thereby prompting an immune response that might raise so many antibodies that some would slip past the blood-brain barrier and incite glial cells to mop up A-beta aggregations. His suggestion drew incredulous stares and even a few chuckles. Vaccines were traditionally employed against viral diseases like polio and the measles. But a vaccine for a noninfectious disease? Attempted in mice? Intended for the human brain? What a long shot! One argument was that not enough antibodies would make it across the blood-brain barrier. Schenk countered by citing published evidence that at least small amounts of antibodies got past and into the brain in both mouse and human.

As far-fetched as Schenk's plan seemed, the others agreed to give it a try. They had so little faith in it, recounts Schenk, that the thinking was that, at the very least, something might be gleaned from the animals' deterioration. Later that day, when Schenk brought his proposal back to his labmates, they proceeded to post it in a corner of the lab's bulletin board reserved for outrageous fantasies.

Given low priority, the vaccine approach was only just getting off the ground in early 1997, according to Schenk. The protocol involved injecting aggregated A-beta 42 peptides once a month into the hind muscle of very young as well as middle-aged mice genetically prodded to produce high levels of A-beta. It would take several months to see if the vaccine had any effect. Schenk's thinking was that if by some miracle it worked, it might keep amyloid aggregates from forming and perhaps even break down already existing deposits. Although many of us believed that the plaques were benign, no one was entirely sure. Moreover, shrinking them might reduce the inflammatory response that possibly heightened neuronal death.

"The really big issue we were facing was whether the clearance of amyloid plaques would do the trick in treating symptoms," recounts

Schenk. "That we were so lucky to even be able to be asking and testing this question! Without risk there's no gain."

If one had a solid theory about how to restrict A-beta's accumulation, and the resources, it was too tempting not to give it a try. At Mass General, Ashley Bush and I were pursuing our own alternate approach. Our impetus was Bush's previously made observation that zinc, copper, and iron can rapidly induce A-beta's conversion into amyloid. High concentrations of all three of these metals in Alzheimer plaques were being reported on by others. In our own tests, it looked as though copper and iron were all the worse in that they sped up A-beta-related free-radical and oxidative stress.

What if chelating agents that lift metals out of solution could be put into a pill and sent to the brain? Would it suck metals out of clustering A-beta and amyloid plaques, dissolving aggregates as well as curbing the production of free radicals? Bush and I elected to test this premise, and Bush's lab on the second floor of Building 149 set about the inquiry. Several chelators worked quite nicely; they prevented A-beta's clumping in the test tube, and some even dissolved amyloid taken from autopsied Alzheimer brains. Yet chelators soon struck us as too harsh a treatment. One risked removing too much metal from the body, robbing it of vital amounts. (Such is their strength that when chelators are added to certain foods, it ensures that metal-dependent types of bacteria won't survive.) We turned, instead, to a less voracious metal-binding agent, a once-prescribed compound. It had been taken off the market due to adverse side effects in a small group of Japanese, a drawback we hoped could be corrected. Later in '97, with backing from two Australian venture capitalists, we would incorporate the company Prana Biotechnology solely for the purpose of developing new drugs based on our compound's potential as a metal-ousting anti-amyloid therapy.

Over a decade earlier, the big push had centered around finding the first gene mutations. Here it was, a brief hop forward in time, and the field was headlong into contriving a wide assortment of drug approaches. We didn't have Alzheimer's figured out by any means, but explaining a disease's mechanism in toto isn't necessarily the fastest way to a drug. With respect to our Baptist contingent, George Glenner's A-beta peptide had launched us, and now, having come full circle, once more we were riveted to A-beta. We'd come so far so fast that even as drugs were in the throes of invention, at the other end of the research spectrum still hung the *what* question. *What* and *where* were other genes that contributed to late Alzheimer's?

This inquiry remained the predominant focus of many Alzheimer genetics teams, ours included. In Building 149, our data from the NIMH initiative's 300-plus Alzheimer families continued to hang steady. Chromosome 3 still signaled something amiss. Yet by mid-1997 most of Marilyn Albert's and my joint efforts were redirected toward chromosome 12. Signs of a late-onset risk gene had lured us to the chromosome's short arm, where we were circling A2M, a gene that was fetching the attention of other investigators. Alpha2-macroglobulin, the protein it encoded, had been found in the peripheral swirl of amyloid plaques. Moreover, in its normal state, after being released by a cell it was known to reenter a cell through the same membrane door, or receptor, used by both the APP and apoE proteins. So conceivably this door and its binding proteins spelled trouble in Alzheimer's. One variant of the A2M gene was known to contain a five-base deletion. Wondering whether it might be Alzheimer-related, we started subjecting this variant to genetic-linkage analysis. Yet our tests kept coming up neutral, and toward the end of '97 A2M still looked innocent.

That October, Peggy Pericak-Vance's Duke team, whose chromosome 12 work had caught me off-guard at the Osaka meeting, published their chromosome 12 evidence in the *Journal of the American Medical Association*. Their paper was widely discussed and raised plentiful queries from the press. When I got hold of a copy, what I saw made me concerned about our own chromosome 12 candidate—A2M—because the Duke team's site of interest was so distant from ours. It was near chromosome 12's centromere, millions of bases to the south of A2M. When we applied additional markers to carefully analyze their region, our results were entirely negative.

Try as we might, we couldn't squeeze out any hard evidence that A2M was an Alzheimer susceptibility gene. We were realizing that traditional linkage analysis simply might not be a powerful enough informer of a late-onset gene, given that we lacked sufficient parental DNA with which to visualize a pattern of inheritance and had to rely on DNA largely from siblings.

In the winter of 1997–98, however, we got a windfall. It came about thanks to a new statistical analysis approach devised at Harvard's School of Public Health by Steven Horvath, a doctoral student, and his adviser Nan Laird, then chair of the Department of Biostatistics. Deborah Blacker in Albert's lab served as a valuable go-between, bringing it to our attention. Under Blacker's and Laird's supervision, Marsha Wilcox, a postdoc in the school's epidemiology department, had used this novel statistical yardstick to compare unaffected siblings from the NIMH families with their affected siblings to see if the inheritance of the A2M variant was tied to the disease's onset. The results, recounts Blacker, turned out to be "wildly significant." The sibs with the deletion-containing A2M variant were roughly three times more likely to have Alzheimer's than those who had a common variant of A2M. I was amazed. How could this novel analysis be generating such different results from our previous assessments? All but simultaneously, a test at Penn bore out similar data.

One doesn't know at this still-early stage of research into disease-prone susceptibility genes exactly what to expect of them. In the case of A2M, in some people it might be sitting on the sidelines not contributing to the disease, while in others it might be playing quarterback in league with other coinherited risk genes and/or environmental factors. My own best guess is that when A2M is a player, it may be altering A-beta's degradation and clearance in the brain, allowing A-beta to accumulate to pathological levels.

This touches upon just about the most important revelation we've so far grasped about the genetics of Alzheimer's. Normally as you age, brain levels of A-beta appear to be maintained by a balance between A-beta's production and its clearance. Early-onset mutations disrupt that balance by turning up the production of A-beta, especially the 42 variety. Late-onset genes seem to do the devil's handiwork by impeding the rate of A-beta's clearance and degradation. Thus we can try to develop drugs that either block A-beta's production or accelerate its breakdown in the brain.

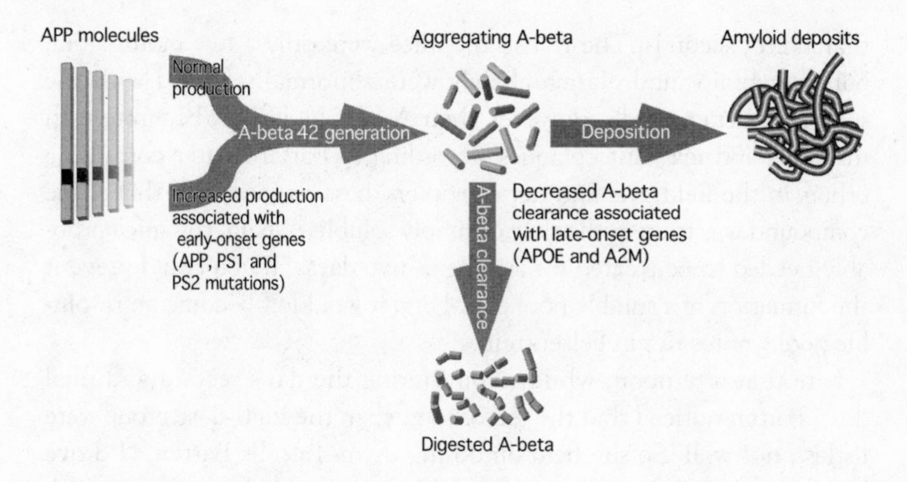

FIGURE 12.2 Increased production of A-beta 42 versus decreased clearance of A-beta 42. Illustration: Robert D. Moir

While our chromosome 12 locus was gaining strength, in the fall of 1997 Bristol-Myers Squibb and SIBIA were making heartening progress. Chemists in both houses had their eyes fastened on a series of gamma-secretase compounds related to the compound fine-tuned at SIBIA. It had worked to dramatically lower the levels of A-beta in cell cultures, and Bristol and SIBIA scientists had begun administering it to transgenic mice. So far it appeared free of major side effects. Yet in the Wallingford lab of Donna Barten, a senior research investigator at Bristol who was directing the mouse tests, the compound was only mildly inhibiting the animals' high A-beta levels. And at SIBIA, months of treating mice with the compound had had no effect on A-beta whatsoever.

"People were getting frustrated," recalls Maria Kounnas. "The chemists had their ideas—it was a problem with the animals. The in-vivo biologists had their ideas—it was a problem with the compound, the speculation being that it wasn't getting into the brain."

Very early on the morning of October 17, 1997, Donna Barten started in on her appointed rounds, dosing her lab's flock of multicolored mice with yet a newer version of the compound. Inserting a tube in a mouse's mouth and injecting a few milligrams of the formula took no longer than thirty seconds. The transgenic mice were only a few months old, but their brains and plasma already were abnormally loaded with the soluble A-beta peptide. However, their A-beta hadn't yet begun to turn insoluble and mass into plaques, according to Barten. After consulting others in the field, she and her coworkers' best estimate was that if the compound was to stem the flow of simply soluble A-beta, the mice probably needed to be treated for as long as five days. "If you could prevent the formation of a soluble pool of A-beta, it wouldn't become an insoluble pool," notes Kevin Felsenstein.

Late that afternoon, while administering the day's second and final dose, Barten noticed that the ten or so mice in the high-dose group were listless, not well. So she held off dosing them. Recalls Barten, "I drove home—an hour's drive—gave my children supper and put them to bed. But I was worried about the mice, and since my husband was home to keep an eye on things, I decided to drive back to the lab to check on the mice. A group of animals like that are important to us. It takes a long time to breed them." Back at the lab, the animals' condition hadn't improved, and Barten made the decision to withdraw them from further testing. Tissue samples were removed from their brains and subsequently analyzed. The upshot, relates Kevin Felsenstein, was "a totally unexpected result." "We saw an 80 percent reduction in soluble A-beta compared to our control animals."

Recalls Barten, "I was shocked, and everyone else was too. We'd been thinking that A-beta accumulates over time, and that a week's worth of dosing was a good time frame for testing, and then to see such a dramatic reduction of the protein after only thirteen hours!" Not only was the compound working in mice, but only one dose did the trick, and in a very short period. "That was a turning point. That compound, though it showed toxicity, would lead to other versions, accelerating our whole program," says Barten, who felt a deep emotional tug because of the progress. An aunt she was very close to had Alzheimer's.

Bristol's first disclosure of the compound's excellent showing in mice was made two months later, in a December 1997 meeting with a group of Wall Street drug analysts. The first announcement to the research community occurred the following May at the 1998 International Business Communications' conference on Alzheimer's disease. Chaired by Dale Schenk and me, it was held that year at Boston's Tremont House

Hotel. There were updates pertaining to a number of promising research fronts, but it was Jeffery Anderson's talk on the second afternoon that particularly provoked curiosity, word having circled that Anderson—a pharmacologist in charge of mouse-testing at SIBIA—had enviable results. No one was out in the hallway drinking coffee and sitting this one out. And during Anderson's talk, no drowsy heads wobbled, no occasional snores erupted, as they had during two previous after-lunch sessions. "Our compound 'A' appears to act as a gamma inhibitor. It has reduced A-beta in mice," Anderson stated. The rest of his talk, which included mention of two other related compounds that hadn't panned out, consisted of a very general sweep of the approach taken, for he wasn't at liberty to say a speck more than what Bristol had disclosed in December. In response to questions asked after his talk, his constant refrain was, "I'm sorry, that's all I can say at this time."

It felt like a historic moment. Whether or not it was wouldn't be known until the researchers learned more about their compound's toxicity, absorption, distribution, and potency, and ultimately tested it in humans.

The best evidence that stopping A-beta would subdue Alzheimer's was, of course, the discovered early-onset mutations. This much was known: these defects generated more A-beta, especially A-beta 42, and people possessing just one such defect were stricken by dementia. Therefore, it seemed, A-beta begot the disease. Or was A-beta simply a strong correlate of the disease?

Countless researchers still didn't think the sticky, aggregable fragment was the whole story. Picture a building ripped apart by an explosion and bomb detectives sifting through the wreckage and separating out what they suspect are tiny pieces of the bomb, then spending years examining these pieces. In the end, they might find that these shards amounted to only one part of the bomb, alone incapable of detonating. Or perhaps they had nothing to do with the bomb, but were just part of the general rubble. Baptist detectives examining the brain's widespread destruction similarly might also be on the wrong trail, thought Tauist types as well as others who bowed down in neither temple. The abnormal tau in tangles, reduced levels of RNA, mistakes in protein synthesis, malfunctioning mitochondria, dangerous levels of calcium in the cell, DNA repair problems—any one of these or other problems might lie closer to the disease's genesis. There were even those who clung to the belief that Alzheimer's was caused by an infectious agent.

Today the skepticism surrounding A-beta has hardly abated. Reflecting the view of many, Nikolaos Robakis at Mount Sinai School of Medicine states, "I don't think amyloid is the whole answer. Many people are going in the wrong direction. Everybody needs something to pin the disease on." In his opinion, since the data isn't clear-cut, room exists for other theories.

Veteran Alzheimer researcher Robert Terry worries over the fact that the amyloid hypothesis is soaking up the field's funding sources while "diminishing the amount of money given to research in other areas." "It's dangerous to be wedded to one hypothesis. Scientists should be more open-minded," says Terry, who is partial to his own theory. More along Tauist lines, Terry believes that the disease's detonation is more closely related to the injury of microtubules inside neurons, which impedes the flow of vital cargo and results in tangles within cells and lost synapses between cells.

No one yet knew for certain that tangles killed neurons, just as it wasn't decisively known if aggregating A-beta did so. Since the early 1990s, scant new clues had been dug up to support a front role for tau and tangles in the disease. Very noticeably, tangle research was in a slump. "There weren't many good investigators—we didn't have enough playmates to play with," mentions Virginia Lee at the University of Pennsylvania. For Lee, John Trojanowski, and others who had vigorously pursued tangles, "it was a gut feeling that tau was important in the disease, though we didn't have as much to hang our hat on," allows Trojanowski.

Tangle scientists had put stock in the observation that tangle amounts correlated more closely with lost neurons and diminished cognition than did amyloid plaques. This suggested that the tangles were the disease's wrongdoer. Yet this argument was losing out to the realization that accumulating A-beta fibrils might start harming cells *before* they amassed into plaques, and *before* tangles appeared. Also, no relevant mutations had been found on the tau gene, so blaming Alzheimer's inheritance on tangles seemed out of reach.

"Then we got a gift from nature," observes researcher Ken Kosik. The cry went out: *Mutations on the tau gene!* The defects, which by June 1998 were being spotted by several teams, didn't account for

Alzheimer's but for early-onset cases of a different dementing disease, which for Tauists was close enough. Sometimes confused with Alzheimer's, the clutch of disorders is collectively known as "fronto-temporal dementia with parkinsonism," or FTDP. Along with dementia, FTDP's diverse symptoms include severe personality changes, language deficits, and Parkinson's-like tremors. Within a year, as many as a dozen FTDP-linked mutations on the tau gene would be discovered, each blamed for lousing up the tau protein and resulting in acutely ill, tangle-choked neurons.

Instantly, tangle research was reinvigorated. "Having worked on tau, we felt vindicated," recounts John Trojanowski. He and Virginia Lee were particularly celebrated for being wedded to an Alzheimer's-tangle connection. "Patients with fronto-temporal dementia don't have amyloid plaques, so it was proof that a tangle disease on its own can kill you. It provided credibility that tau is something toxic and bad"—that abnormal tau very conceivably played a leading role in Alzheimer's.

Jerry Schellenberg's University of Washington team had produced the first report of an FTDP mutation on the tau gene, and Trojanowski recalls an exchange he had with his good friend soon after. "So this breakthrough must be great news for you and Virginia," Schellenberg mentioned to Trojanowski. "At least you won't suffer from 'mutation envy' anymore." Trojanowski was quick to defend. "We never suffered from stupid 'mutation envy'! We knew abnormal tau was bad for the brain and didn't need any tau gene mutations to convince us." There really appears to be such a thing as "mutation envy," and until relieved of it, a geneticist can walk the moors of DNA very dejectedly.

But to many of us in the amyloid camp, the tau mutations failed to excite. Yes, they showed that cognitive loss could be directly caused by the formation of tangles, but there were many sound studies that showed that A-beta starts aggregating and plaques start forming long before tangles appear. This meshed with the theory that A-beta drives Alzheimer's disease, including the tangles. Nevertheless, corrupted tau possibly helped carry out neuronal death. If A-beta is the primary toxic agent—the bullet that lodges in the cell surface and triggers a death mechanism in cells—the resulting abnormal tau inside the cell might be the broken blood vessel caused by the bullet, which results, figuratively speaking, in the cell bleeding to death.

Once again white sails filled the Charles River, fans were doing "the wave" at Fenway Park, and Louisburg Square's grass oval was turning emerald green. The spring of '98 couldn't have arrived soon enough, as far as I was concerned. As mild as Boston's winter had been, it had seemed more prolonged than usual due to the frustrating chasm between our team and the Duke team in regard to our differing positions on chromosome 12. Our MGH group was readying its report on the A2M gene finding, word of which spread like wildfire through the field, prompting various groups to reinspect A2M. But as I suspected might happen, many researchers, employing the traditional case-control method, were unable to detect A2M's culpability in late-onset cases. Right up to the present, it's been my sense that neither standard genetic-linkage nor case-control studies unveil susceptibility genes as effectively and reliably as does the Harvard School of Public Health's family-based statistical model, which serves as a sensitive hybrid of both older approaches.

Nature Genetics accepted our A2M paper for August publication. Publishing the first gene nabbed by such a novel approach felt like a true accomplishment. Before the paper ran, I planned to announce the finding at the Sixth Annual International Conference on Alzheimer's Disease and Related Disorders in Amsterdam—somewhat of an anticlimactic gesture, since so many in the field already knew about it. In mid-July, on the eve of my departure for the Amsterdam meeting, a reporter called and drawled, "What's so important about *this* gene?" She sounded generally fatigued over having to cover yet one more gene finding.

In Amsterdam, while the press generally showed excitement over the evidence I presented for A2M's link to late-onset cases, I experienced—in some circles—a tidal wave of, worse than boredom, defiant incredulity. "Rude Journée Pour Rudy," a French research publication headlined. Rough day for Rudy indeed! Two years earlier at the very same meeting, then in Osaka, I'd announced our linkage to chromosome 12, and Peggy Pericak-Vance had stood up to say her Duke team also was after a gene on 12. Now I was announcing we had the actual gene, and once more the rain cloud opened, Pericak-Vance again rising and this time questioning A2M's validity as an Alzheimer gene. I thought my lab's data coupled with a thorough explanation of our analytical method would settle the A2M issue. But it didn't. Members of the Duke team described their evidence that A2M had minimal bearing on the disease, one scientist—my old Huntington's colleague, Mike Conneally—declaring that "Rudy's hype—, er, hypothesis" was simply without grounds. I made an appeal for more open-mindedness and suggested that my detractors try out the new analysis technique before shooting it down along with the gene.

Covering the Amsterdam fireworks, a *Science News* journalist observed, "It's naive to think that scientists aren't as combative as athletes or business-people. And nothing gets the competitive juices flowing as quickly as the hunt for a gene responsible for a feared human disease."

As drugmakers crafted projectiles that might deter the still-invisible, hypothetical, enigmatic secretases—the liberators of A-beta—other scientists were simply trying to find and identify them. At the fall 1998 Society for Neuroscience meeting in L.A., amid the usual deluge of reports, Dennis Selkoe threw out a lightning bolt. His lab had discovered that the early-onset presenilin 1 protein appeared to function as a protease—a snipper of other proteins. He theorized that presenilin *itself* might be one of the secretases—the gamma-secretase. It had been imagined that when presenilin escorted APP inside the cell, it put APP in the path of the secretases, and that if presenilin was mutant, APP released *more* treacherous A-beta 42. But here was a much cleaner, more logical hypothesis: presenilin, nuzzled against APP, might be directly releasing A-beta. If presenilin 1 was one of the scissors that set loose A-beta, that implicated it in all cases of Alzheimer's, not only early-onset. Yet there was no precedent for a protease that structurally resembles presenilin, and science loves precedents. The theory met up with a fair dose of skepticism, "Hogwash" being one reaction.

All too often lately we were sounding like a busload of bickering old men. The more progress, it seemed, the more kvetching. The greater a finding's potential, the louder the protest.

The L.A. Neuroscience meeting thankfully provided some sorely needed comic relief. Awarded to members of the field who had distinguished themselves in unique ways, the "Alois Awards"—dreamed up by Dale Schenk—were presented by Schenk and me with as much Billy Crystal quicksilver commentary as we could muster. Sealed envelopes and replicas of gold "humanoid" Oscars lent a distinct flavor of L.A.'s swankier limelighted night. "Aloises" went to Sam Sisodia for Best Highlighted Yet Confusing Research Topic (his rather complicated reports on the presenilin proteins); to the Mandelkows, a famous German husband-and-wife tangle team, for Best Resurrected Research Topic (their seminal contributions to tau/tangle research); to Matthias Staufenbiel from Novartis for Boldest Choice for a Transgenic Spokesmodel (he was male; most others of the field's mouse spokesmodels were female); to Dora Games for Most Slides Ever Shown in a 12-Minute Neuroscience Presentation (an impressive sixty); to Dennis Selkoe for Best Persistent Research Topic (the amyloid hypothesis—what else?); to Steven Younkin for Most Thoroughly Beaten Research Topic (A-beta 42—the most studied molecule in Alzheimer's); and to Allen Roses for Person Most Responsible for Consolidating Support for the Amyloid Hypothesis (for his famous disbelief in the amyloid hypothesis). This last presentation especially brought down the house.

I too failed to escape, receiving an Alois for the Best Elaborate Research Topic. Apparently my analysis showing that A2M might be a late-onset gene was considered a bit inventive.

The true Oscars annually bestowed to our crowd are the prestigious Metropolitan Life Foundation Award for Medical Research and the Potamkin Prize for Research in Pick's, Alzheimer's, and Related Diseases. Deservingly, most of our field's leading geneticists had received these tributes over the years. Everyone had done well; at some point everyone had won their own race and gotten their just reward—including George Glenner, who in the late '80s had received both a Met Life and a Potamkin prize. These honors had come my way as well, in '95 and '96 respectively.

As our various contributions had laid a road of progress, the sense had deepened that we were comrades united against a much larger foe, one that grew ever more colossal and complex the closer we got. Despite the times of discord, despite the rivalries, there was the recognition that without our big tribe, each of us would be an ant lost in tumbleweed. Keeping stubborn watch over our own turf and findings, we could behave ridiculously autonomously. But away from the lab, together in far-flung places, so often we put aside our dueling natures and traded shop in the languid-orange glow of a bar in Paris or Berlin, or Zurich, London, Tokyo, Bali, Florence, or Seoul, or on Daydream Island off Australia, all in a comradely atmosphere. For who better understands our obsession with this disease? Who better to talk to than those whose findings have lent to our own? "If I have seen further," wrote Sir Isaac Newton to Robert Hooke three centuries ago, "it is by standing upon the shoulders of Giants." Not that all of us are giants. But you can bet that if someone isn't standing on your shoulders, peering off your latest sightings, you're peering ahead off theirs.

In making his famous remark, Newton himself seems to have stood on another's shoulders, namely the Roman poet Lucan, who in the first century A.D. observed that "Pigmies placed on the shoulders of giants see more than the giants themselves."

At Bristol-Myers Squibb and SIBIA, the compound shown to successfully reduce the A-beta peptide in rodent brain tissue was continuing to undergo modifications, readying it for human trials. Rumor circled that Lilly and Athena's A-beta inhibitor was even further along and perhaps on the very threshold of human testing. Yet not a peep was heard from these collaborators about their prototype.

As we headed into 1999, several happenings furthered the sense that before long, by hook or by crook, the field would have an answer to the question so many of us had been pursuing ever since George Glenner handed us the biochemical key to Alzheimer's plaques—whether the accumulation of A-beta and amyloid in the brain actually was Alzheimer's poison. In my own lab, the news arrived from one of our Australian research partners, Colin Masters, that the metal-binding compound fine-tuned by Ashley Bush had reduced brain amyloid in twelve-month-old transgenic mice by 40 percent, on average. It appeared to dissolve aggregated A-beta. Very excitingly, amyloid had been entirely eliminated in 25 percent of the mice! The therapy showed minimal toxicity, and a second trial with older, more amyloid-burdened mice was immediately designed. The first-round results held out such promise that our Prana partnership decided to go the distance. Plans to proceed toward clinical trials got under way.

In the laboratories of Athena—whose name had changed to Elan Pharmaceuticals—no one was chuckling anymore over the seemingly cockeyed suggestion made by Dale Schenk that a vaccine consisting of A-beta 42 might detain the buildup of amyloid plaques in brain tissue. In July 1999, Elan shared with the rest of the world what it had known for well over a year. In one test, Elan researchers had injected the vaccine into young transgenic mice—mice engineered to process too much A-beta and amyloid—on a monthly basis for a year. At thirteen months, when the mice were analyzed, very few plaques were observed, whereas in untreated mice of the same age plaques flourished.

"We saw almost nothing. *Nothing*," recounts Schenk. Hardly any amyloid deposits. "I was unbelievably surprised. You think, in this type situation, that the positive data is going to go away." But it stayed. Recalls Ivan Lieberburg, then director of research and currently Elan's chief scientific and medical officer, "When I saw the original results, basically I was shocked. I thought we had made a mistake and that we'd treated control animals"—mice not engineered to develop plaques. But no mistake had been made.

Jaws dropped even more at Elan. In a second test, when the vaccine was administered to one-year-old mice that already had plaques, it curtailed the formation of new plaques. In some instances it even appeared to break down and get rid of existing plaques!

These hopeful mouse tidings swept across newspaper, television, radio, and the Internet. For Ashley Bush and myself, it was good news for our own less-publicized mouse work, because it was further testimony that plaques could be removed. Our compound stripped metals from congealing A-beta, reducing fibrillar A-beta's ability to form plaques while helping to dissolve existing plaques. Elan's vaccine, in comparison, raised antibodies that recruited massive amounts of microglia—more than plaques attracted on their own—that ingested fibrillar A-beta and amyloid and got rid of it. Both approaches implied that a therapy could not only stall amyloid formation, but undo existing amyloid deposits. The tremendous suggestion thrown forth, then, was that Alzheimer's might be reversible.

Elan's was the first published report to show this. If—and it remained a very large *if*—A-beta and amyloid had toxic properties, it followed that preventing and reversing amyloid pathology might be a treatment path not only for patients who were newly symptomatic, but even, perhaps, for those in advanced stages. The removal of aggregated A-beta possibly could allow for the regeneration of the neural circuitry, just as stopping smoking can pave the way for healing lungs.

Unavoidably, our metal-extracting drug as well as Elan's vaccine raise certain concerns. In our case, will impeding fibrillar A-beta and dissolving amyloid cause freed A-beta to harm the brain? In Elan's case, plaque amyloid is being consumed by microglia; but will microglia be able to fully digest and terminate A-beta 42? (Other studies, like those of Charles Glabe, a cell biologist at UC Irvine, indicate that A-beta 42 actually remains intact inside microglia.) Elan's vaccine poses other potential problems. Some still question whether enough antibodies can get into the brain to have a decent effect, since the human blood-brain barrier is more discriminating than that of the mice the vaccine has been tested on. If enough do, there's the risk of an autoimmune response—too many activated glia attacking brain tissue and good proteins. Autoimmune problems might extend to nonbrain tissue.

Ever since the transgenic attempts in the early '90s, plaque-bearing mice that could demonstrate a drug's effectiveness in turning down amyloid had been the holy grail. And here was a row of superlative successes! But the distance from mouse to human is a huge leap for a drug, and many don't make it. It isn't just that a mouse brain weighs one gram, a human brain some 1,400 grams. The former contains a hippocampus, cortex, and other structures seen in humans, but "it is wired very differently," notes Ivan Lieberburg at Elan. "Many drugs that work in mice and rats show no activity in humans."

Critics of transgenic mice have long asked, How can a brain of even an elderly mouse begin to mirror the developed complexity of a seventy-year-old human brain? And how can one assume that a transgenic mouse brain makes human A-beta and amyloid the same way a human brain does, or, if treated, responds the same way? And it didn't even look as though A-beta in mice killed many neurons—which could make one a bit queasy about the amyloid hypothesis. Was it because mice, even in response to A-beta's toxicity, are incapable of making tangles, and that tangles are the coffin nails in neuronal death?

The true test would have to be in humans. Elan's A-beta 42 vaccine would begin clinical trials in patients in December 1999; the Bristol/SIBIA A-beta inhibitor would follow suit in March 2000. Our company's metal-reducing A-beta therapy, meanwhile, was being evaluated for human trials. Once a drug entered the clinical forum, it would probably take at least two or more years to judge its safety and effectiveness. What a special start to the millennium it might prove to be. Or what a major disappointment.

Right before the century ended, a final gust of superb news blew in. In late October, Robert Vassar, Martin Citron, and their teammates at Amgen, a biotech establishment in Thousand Oaks, California, reported in *Science* that they had found none other than one of the two elusive secretases that cut A-beta loose—the beta-secretase, gamma-secretase's fellow conspirator. There'd been so many unsubstantiated claims of trapped secretases that instinctively many of us mistrusted the account. Explaining his own initial disbelief, Sam Sisodia told the *New York Times* in a front-page story, "It has been junk after junk for twelve years." The Californians, however, appeared to truly have the real thing. After a two-year search, they had retrieved it in a brute-force way that only a company with expensive high-throughput equipment could. A secretase being a protein, they'd looked for its gene by inserting random genes into cells and taking notice of those whose protein made cells churn out more A-beta, distilling the pool down to one prime suspect. With the beta-secretase's structure in hand, no longer would drugmakers be forced to design an A-beta blocker blindfolded.

But no one is ever alone. Amgen's announcement brought out of the woodwork three other companies who'd also snagged beta-secretase—SmithKline Beecham, Pharmacia-UpJohn, and Elan Pharmaceuticals, the vaccine-maker. The upshot, as it looms today, could well be a prolonged, noisy battle over the secretase's patent, which likely will go to whichever company can prove it found the enzyme first. This won't, however, keep these companies from rapidly pursuing compounds aimed at beta-secretase.

Where we'd arrived was so reminiscent of the gains made in AIDS research in the mid-1980s. Researchers, once realizing that the HIV virus's replication depended on the cutting action of a protease, proceeded to isolate that enzyme and acquire the means for developing protease inhibitors, which have given so many AIDS patients a new lease on life. On the Alzheimer front, however, the secretase-based A-beta blockers were advancing within the tentative confines of the amyloid hypothesis, A-beta's guilt not yet a certainty. Yet here was a secretase, finally proven real.

There's a small seaside restaurant on Boston's South Shore where I like to go for brunch on Sunday mornings because of the salt kick in the air and the fact they serve a mean eggs Benedict on avocado. A friend and I were there not long ago when I noticed a frail, white-headed man sitting at a nearby table with his family. Those around him were caught up in an animated discussion, while he sat silently wedged between them. The look on his face was vacant and flat, so like the hollowness commonly seen in people with Alzheimer's.

Suddenly there was a lot of commotion around his table. While eating, the same gentleman apparently had swallowed wrong—another sign that possibly he had Alzheimer's—and was choking. A waitress had run over and, clasping him from behind, was attempting the Heimlich maneuver. As much as a minute probably went by. Her efforts weren't working, and a big guy wearing a Patriots' jacket took her place, but to no avail. Color was rapidly draining from the man's face, so I ran over to try to help out. All but holding the fellow up, I proceeded to give him such a hard series of squeezes that I'd later worry I'd broken some of his ribs. On the fourth squeeze a piece of sausage literally shot out of him. Someone had called 911, and by the time the fire truck rolled up, the man had begun to revive and regain color. He and his family were greatly appreciative, and a few weeks later the town's fire department sent me a letter of gratitude.

How ironic it all seemed. To be miles from the lab and at long last perhaps have finally saved the life of an Alzheimer patient, if only temporarily. The Way of Things.

The Way of Things had, as well, brought me into contact with the Noonan family. A few years earlier we had appeared on *20/20* together in a segment on Alzheimer's disease, but we had been filmed separately for the show and hadn't met. Having decided not to sit by passively as the disease clutched certain among them, the family had not only provided Mass General with blood and skin samples for DNA probing, they also were sharing their experiences with groups large and small and had started an annual bike ride to rally support against Alzheimer's. I had said I'd try to bring them up-to-date about where the science was headed. It was the first time I'd gotten together with a family whose DNA samples—all anonymously numbered—my unit worked on.

And so one day eight of Julia's ten children and their spouses met with me in Building 149's sixth-floor conference room. They gave the impression of a perceptive, energetic group of very close friends, many around my own age, who were eager for knowledge that could lend to their brave fight. Telling them I felt I was preaching to the choir, I reviewed what we knew of the genetics and pathology. They already had a pretty solid sense of the molecular picture I sketched out.

As I turned the corner into potential avenues for therapies, I found myself, as I always do when speaking with those at risk, walking a precarious fine line. I wanted them to share in the research community's optimism, many of us being of the opinion that an effective treatment might be available in five to ten years. But I didn't want to give false hope. After all, crucial pieces of the puzzle were still missing: the precise mechanism through which the early-onset mutations cause the disease; why A-beta builds up in just the brain when it's released bodywide; to what extent A-beta and tau are interrelated; and whether it's really true that A-beta is primarily responsible for Alzheimer's devastating loss of neurons. The devil lay in the details, and there were a tremendous number more to pin down. But the progress had been exponential. For every molecular action sighted, science had a drug target, and over a dozen such targets were now visible. A reported sixty-plus new Alzheimer drugs were in the throes of development.

It was impossible not to have expectations; after all, we stood on the brink of captivating drug trials. There was light to steer by, and science's fondest findings—a string of hard-won truths—carrying us forward.

August 1999. Twenty-one bicyclists depart the Berkshires and in the course of two days, under cloudy skies, sometimes skimming through puddles, they pedal clear across Massachusetts, cutting a swath through a state where 130,000 people are afflicted by dementia. With each mile, "Memory Ride"—launched three years earlier by Julia's children—draws hope, awareness, and research dollars for Alzheimer's disease. This year's participants will raise almost $50,000—nearly twice last year's total and enough to cover the salary and supplies for one junior researcher for an entire year. Many of the riders have had parents taken by the disease; some among them have early symptoms. As they lean up Beacon Street and end their ride in front of Boston's gold dome, they are but a small group "chasing a shadow"—one of Julia's daughters' descriptions of her illness. The assemblage—riders, crew, Julia's children and grandchildren, and well-wishers—carry a message pure and simple into the State House's Great Hall: "We don't want our children to know this darkness. We can conquer it."

Epilogue

As this book travels to press, biology's Big Bend has been rounded. Scientists from Celera Genomics and the federally funded Human Genome Project have jointly announced, well ahead of time, that they've all but decoded humankind's DNA and its letters of life. It was only in 1995 that science read the first genome of an independent organism—a bacterium named *Hemophilus influenzae* (1,743 genes). Yeast followed in 1996, the first cracked code of an organism with a cell nucleus and therefore kin to all plants and animals (6,000 genes). Two years later, scientists unveiled the entire sequence of the first multicellular organism—*C. elegans*, the minute dirt worm (19,099 genes).

And now *Homo sapiens*. At last human genetics will have its own Table of Elements for reference. But it's early in the day, with lots more to do. We still need to pick out and make plain the vast numbers of genes within the genome's sequence. And left to identify are the vast numbers of proteins made by genes, their various functions in the body, and, if flawed, the consequences. This could take the better part of the century, although genetic knowledge of other organisms should help out. For instance, each gene—hence protein—in yeast exists in like form in humans.

Let's just hope that as we lean on kindred genomes for insight into our own, we'll gain greater respect for all life. We and the mouse, after all, have genomes that are roughly 97 percent similar. And let's hope that this broader sensibility will guide us in how we will apply our mountain of new knowledge. As Jim Gusella frequently notes, "All knowledge is good; the only concern is how it's used."

As for those human genes connected to Alzheimer's, new reports support our Mass General team's belief that the A2M gene is a late-onset culprit, a run of case-control studies having failed to consistently find such evidence. Three teams adopted the novel family-based technique that led us to A2M and confirmed a correlation between A2M variants and an increased disease risk. Three other groups have tied A2M to increased brain amyloid. At Mass General, meanwhile, the large NIMH DNA screening is revealing several hits for other late-onset genes, with one very strong candidate on chromosome 10. A published report from Ellen Wijsman's lab at the University of Washington suggests that, altogether, five to six major aberrant genes may be involved in late Alzheimer's.

Despite sharper statistical tools, geneticists' biggest problem continues to be securing total proof of a late-onset gene's unruliness, and how it joins with other internal and external factors to bring on the disease. With over a dozen Alzheimer candidates being scrutinized by various labs, and undoubtedly numerous others to be hauled into view, it could take many years to achieve a full accounting. Similar to the genes already cornered, many of us expect that those left to identify somehow inveigle A-beta's toxic buildup.

In mid-1999, Blas Frangione's NYU lab isolated yet another dementia-associated amyloid protein, one that in my view provides an intriguing parallel to what might be happening in Alzheimer's disease. In some ways, Familial British Dementia is markedly different from Alzheimer's; it's rare and, like several other amyloidoses, mostly confined to the brain's blood vessels. Patients' gray matter contains tangles and some dense amyloid plaques, but many more diffuse plaques. A chief similarity, however, is that a gene mutation (on chromosome 13) translates into high levels of an insoluble fibril that free-floats between gray-matter neurons. Distinct from, yet comparable to, Alzheimer's A-beta, this fibril just might be driving neuronal death. It would appear that all it takes is copious, accumulating amyloid fibrils of any type in memory regions to trigger inflammation, tangles, neuronal loss, and dementia. As to why, in this rare disease, dense plaques don't overrun gray matter, the nature of its fibril may not allow for their formation.

Making for rapt discussion among neuroscientists is the fairly new realization of just how many brain disorders follow the same course—from gene mutation to a mucked-up, misfolded protein that fibrilizes and forms rock-hard aggregates the body can't get rid of. Along with Alzheimer's, examples of other maladies that give rise to brain deposits of one form or another—in the following cases, usually inside the cell—include Parkinson's disease, which promotes tiny clusters called Lewy bodies; the prion diseases and their bunching prion rods; Lou Gehrig's disease and its multiple inclusions; and Huntington's and other neuro-disorders that spring from overly repetitious DNA and load brain cells with insoluble particles.

"It's very exciting to see these commonalities emerging," notes Anne Young, Mass General's chief of Neurology Service. "In none of these disorders have protein fragments or their aggregates been proven to be the sole source of the disease, but in several cases the fragments are very likely toxic. Many of these diseases develop slowly, so there's all the more hope of interfering with their aggregating proteins." As for Alzheimer's, it's lately been noticed that aside from its profuse extracellular A-beta, A-beta can accumulate *inside* neurons. Although cell amounts are minuscule, intensive research has begun to determine whether they are harmful.

Several fresh findings indeed square with the conviction that amyloid fibrils are a driving component of Alzheimer's pathological stream.

To mention but a few, Mike Mullan and coworkers have seen strong evidence in mice that the A-beta peptide activates the immune system's microglia, which, as we've described, may contribute to the demise of neurons. The University of South Florida team further has shown that if you keep microglia from reacting to A-beta's presence, you might reduce neuronal loss, which suggests yet another drug target to pursue.

For all those who thought "hogwash" when Dennis Selkoe's lab proposed that Alzheimer's presenilin 1 protein might itself be the gamma-secretase, one of A-beta's liberating proteases, lately both a Merck team and Selkoe's lab independently lassoed more direct evidence. That PS1 may be the gamma-secretase demands more proof, but if it stands up, just as drugmakers are doing with the recently nabbed beta-secretase, compounds might be made to fly in the face of presenilin with the intent of reducing A-beta fibrils.

Two studies lend further credibility to the idea that accumulating A-beta in the brain is Alzheimer's weapon of mass destruction. Bruce Yankner's group at Children's Hospital discovered that injected A-beta has a toxic effect on the brains of older monkeys, but not younger primates or mice. This would seem to indicate the extent to which aging acts as a susceptibility factor in Alzheimer's, and that A-beta's toxicity is specific not only to older brains but to species with more highly developed brains. It's tempting to think this might explain why transgenic mouse models overloaded with A-beta fail to exhibit substantial neuronal loss.

Finally, a study by Joseph Buxbaum's team at Mount Sinai and collaborators at Rockefeller University and Albert Einstein has shown a strong correlation between total elevated A-beta and the progression of dementia. It cannot be concluded that A-beta is forcing the insidious onset of dementia. Yet if there were any doubts regarding a correlation between mounting brain A-beta and mental decline, this study should do away with them.

On the drug front, one of the most hopeful recent findings was made by Karen Duff's NYU group. After demonstrating that a high-fat diet increases A-beta's production in male transgenic mice, the researchers went on to dose the mice with a cholesterol-lowering drug and, so promisingly, saw A-beta levels diminish. Using the same type drug, Konrad Beyreuther's lab at the University of Heidelberg observed a reduction of normal A-beta levels in guinea pigs. Should too much A-beta in the brain be Alzheimer's fatal flaw, both experiments point to a treatment that's readily available. Precisely how cholesterol-lowering drugs work against A-beta remains a mystery.

In January 2000, my colleague Ashley Bush e-mailed glad tidings from Australia. "Holy Fenoki! Could it be???" his message exclaimed. In a second set of tests, our company's metal-extracting compound had reduced total fibrillar A-beta in the brains of aged twenty-one-month-old transgenic mice by about 50 percent. (In transgenic mice of this age, it's usually hard to find a part of the cortex that isn't overwhelmed with plaques.) Not expecting anywhere near this magnitude of effect, I was totally surprised. Our previous positive results in younger mice just might have been a fluke. The company is currently proceeding to phase one of clinical trials to assess the drug's safety. Once we see safety, we greatly look forward to phase two and testing the compound's efficacy in patients.

With plentiful signs indicating that useful Alzheimer drugs may be on the way, increased attention is being focused on developing sound diagnostic tools. Good diagnostics are critical if, for the sake of preventative measures, we hope to catch sight of Alzheimer's earliest inner changes years ahead of its obvious symptoms. As Zaven Khachaturian has pointed out, while Alzheimer's mostly appears later in life, it is actually a younger person's disease, since its abnormalities fester years before they become manifest. The most beneficial diagnostics will be those that can pick out the disease's genetic faults in a person's genome; or detect its earliest biochemical alterations in the blood and/or spinal fluid; or spot structural changes in the brain.

In the early-onset category, predictive blood tests currently exist for some of the discovered early-onset mutations. Commercially available, they are performed within the framework of a larger neurological examination in many academic medical settings. Each of these tests requires genetic and psychological counseling as well as legal safeguards that ensure confidentiality on the part of the center administering the exam. Some at-risk individuals who have a parent with early Alzheimer's, and who might be experiencing memory loss themselves, want to be tested, for it can help them and their families make important decisions about the future. But since no effective drugs are yet available, most of those at risk very understandably would rather not know.

In the late-onset category, blood tests exist for the APOE variant. But, as already mentioned, while APOE-4 can boost the risk for Alzheimer's, it does not guarantee the disease. Therefore, in most situations APOE testing is not recommended.

Aside from genetic testing, several other diagnostic approaches are being developed. One of the most promising is MRI—magnetic resonance imaging. When used to scan the brains of people who are either genetically prone to or already displaying dementia, MRI can monitor whether regions central to memory—the entorhinal cortex or hippocampus, in particular—are undergoing shrinkage indicative of Alzheimer's. Another use of MRI currently under development will allow doctors to view plaques and tangles directly. Even if these lesions are late-stage phenomena, mounting numbers might provide an early warning. When effective drugs arrive, MRI also might serve to monitor a drug's ability to slow plaque-tangle formation and the brain's volume loss.

In April 2000, a study by Marilyn Albert's Mass General crew, in which my lab collaborated, showed that MRI, when trained on changes in the entorhinal cortex of patients with mild forms of forgetfulness, can indicate how soon a patient will convert to full-blown Alzheimer's. Researchers had guessed the entorhinal cortex degenerated early on. Now to actually *see* direct evidence, to be able to actually watch the disease progress in a living person, is quite amazing. It gives us a way of tracking the rate and severity of the disease's progression; of knowing, for the sake of applying drugs, who will get the disease sooner than later.

There continue to be high hopes that someday, by measuring a specific altered protein in the body, doctors will have a marker that easily reveals whether someone harbors a susceptibility to Alzheimer's. To that end, investigators continue to gauge whether a blood or spinal fluid test can determine elevated A-beta 42 or abnormal tau.

In the future, the ultimate indispensable diagnostic might be a person's genotypic profile. Someone's entire genome might be put on a biochip that is outfitted with all the mutations and polymorphisms known to cause Alzheimer's. Their DNA would either bind or not bind to those danger points. A computer scan of the chip would handily reveal any DNA defects, doing away with the time-consuming methods that currently allow us to assess a person's gene alleles. The resulting profile might relate other valuable pieces of information, such as a person's overall risk for Alzheimer's (depending on which aberrations are present); the time window of onset; and which treatments will be most effective and which could carry adverse side effects.

A futuristic vision for conquering Alzheimer's wraps around the powerful combination of genetic screening—always accompanied by counseling and stringent safeguards—and preventative drugs optimized for a person's genome. A foundation is being laid right now for such an eventuality. The year might be 2010 or later. A person with a family history of Alzheimer's, or perhaps no family history, decides to be assessed for his or her predisposition to Alzheimer's. Family history is reviewed; a genotypic profile is obtained; levels of A-beta, tau, and perhaps other relevant proteins are measured. A baseline measurement of certain brain regions is retrieved through MRI. Should this workup sound an alarm, were just one magic bullet available—a vaccine, for instance—that would obviously be the therapy of choice. But more likely a range of treatments would be required and dispensed according to a person's age, genetic profile, and disease stage. Should fibrillar A-beta turn out to be Alzheimer's demon, a daily A-beta cocktail might be prescribed that thwarts A-beta in a number of ways—by inhibiting the secretases' activity, by blocking A-beta aggregation, and by promoting its clearance. Other drugs could assuage the menacing biochemical features in the disease's cascade or promote neuronal regeneration. Short of curing Alzheimer's, the goal would be to make it a manageable chronic condition.

Currently, there's a growing awareness that the last thing we want to do, in terms of treatment, is to prolong late-stage disease, suspending a patient in a terrible state of nonidentity. An important clinical decision will be at what stage it may be too late to treat someone. Equally important will be the decision of how early to begin treatment. Since it appears that the disease's molecular wrongs commence relatively early in life, there's reason to think that intervention might start at a fairly early age. This seems a small price to pay were it possible to silence Alzheimer's.

By mid-century, new estimates warn, the world may contain 45 million Alzheimer patients—three times as many as there are now. Bring on effective drugs, and a good many of those individuals just might be able to experience their golden years with a clear mind and sound recollection.

Resources

Alzheimer Disease: The Changing Scene
By Robert Katzman and Katherine Bick, 387 pp, Academic Press, San Diego, CA, 2000.
As author Katherine Bick describes, this recently-published book "analyzes the Alzheimer's research environment from 1960-1980, a time when the foundations for the explosive growth of interest in Alzheimer's were being laid. Interviews with pioneers of that formative period provide an unequalled view of the passions and arguments that drove the biomedical researchers, care providers, and family members to attack the seemingly intractable issues posed by Alzheimer's."

Alzheimer Research Forum
website: http://www.alzforum.org
Primarily dedicated to making research findings easily available for professionals, the Alzheimer Research Forum's web site nevertheless is filled with material that non-scientists will find informative. Along with the latest research reports, it includes online interviews with scientists, "Milestone" papers, a treatment guide, a listing of drugs in clinical trials, and helpful links for caregiving.

Alzheimer's Association
919 North Michigan Avenue, Suite 1100
Chicago, IL 60611-1676
(800) 272-3900
website: http://www.alz.org
E-mail: info@alz.org
The goal of the Alzheimer's Association is to eliminate Alzheimer's through the advancement of research while enhancing care through support services for individuals and their families. It represents the country's largest private funder of Alzheimer research and offers widespread support and educational programs through a network of 200 chapters. For your local chapter, contact the 800 number above.

Alzheimer's Disease Education and Referral Center (ADEAR)
P.O. Box 8250
Silver Spring, Maryland 20907-8250
(800) 438-4380
website: http://www.alzheimers.org
E-mail: adear@alzheimers.org
A service of the NIH's National Institute on Aging, ADEAR provides an array of information on Alzheimer's disease, its impact on families and health professionals, ongoing research and clinical trials, and relevant publications.

The George Glenner Alzheimer's Family Centers
San Diego, CA
(619) 543-4700
website: http://www.alzheimerhelp.org
This helpful web site provides information about the George Glenner Alzheimer's Family Centers and the George G. Glenner School of Dementia Care as well as support-group events in the San Diego vicinity.

healthfinder
website: http://www.healthfinder.gov
Developed by the U.S. Department of Health and Human Services, healthfinder is a reliable, user-friendly gateway to news and information on most every health topic imaginable. In addition to links to online journals, medical dictionaries, and support groups, its contains an informative swath of material pertaining to Alzheimer's disease.

Memory Ride
website: http://www.memoryride.org
Memory Ride is a nonprofit Massachusetts cycling group dedicated to fighting Alzheimer's disease. Its participants raise awareness toward the disease as well as funds for Alzheimer's research.

National Family Caregivers Association
(800) 896-3650
website: http://www.nfcacares.org
E-mail: info@nfcacares.org
NFCA, a national charitable organization dedicated to servicing family caregivers, provides infinite resources for those who find themselves in a caregiving role.

Notes

Chapter 1

Page

7 The idea of using DNA variations to map the human genome was just entering the collective consciousness. A now historic paper published that May theoretically set forth the inestimable value of this undertaking. Among its authors were David Botstein at MIT and Raymond White at the University of Massachusetts, Worcester, the earliest and most vocal advocates of making use of DNA variations to dissect the human genome. The paper's reference follows: Botstein, D., White, R. L., Skolnick, M., and Davis, R. W., "Construction of a Genetic Linkage Map in Man Using Restriction Fragment Length Polymorphisms," *American Journal of Human Genetics* 32, 314–331 (1980).

7 Throughout history, Huntington's contortions were frequently and tragically misinterpreted. Various documents provide evidence that in Colonial times a strain of Huntington's connected to an English family that arrived in Boston Bay in 1630 was responsible for certain descendants in that family being viewed and tried as witches—notably a woman named Elizabeth Knapp, the alleged Groton witch. One interesting reference is a paper by P. R. Vessie, "On the Transmission of Huntington's Chorea for 300 Years," *Journal of Nervous and Mental Disease* 76 (December 1932), 553.

17 The G8 probe's general location on chromosome 4 was detected by Susan Naylor, a molecular geneticist at the Roswell Park Memorial Institute in Buffalo. Naylor had developed a unique method for matching bits of DNA to their chromosomal home that involved propagating human chromosomes inside rodent cells.

Study

The first study to localize a disease gene (Huntington's) using random DNA markers:

Gusella, J. F., Wexler, N. S., Conneally, P. M., Naylor, S. L., Anderson, M. A., Tanzi, R. E., Watkins, P. C., Ottina, K., Wallace, M. R., Sakaguchi, A. Y., Young, A. B., Shoulson, I., Bonilla, E., Martin, J. B., "A Polymorphic DNA Marker Genetically Linked to Huntington's Disease," *Nature* 306, 234–238 (1983).

Chapter 2

25 Rudolf Virchow's quote, "Only when we have discovered . . ." appears in a footnote in a chapter on amyloid in Virchow's *Cellular Pathology as Based upon Physiological and Pathological Histology; Twenty Lectures* (New York: R. M. DeWitt, 1858), p. 415.

27 The translation of the clinical description of Auguste D. is excerpted from: *The Early Story of Alzheimer's Disease*, edited by Katherine Bick, Luigi Amaducci, Giancarlo Pepeu (Padova, Italy: Liviana Press, 1987). Distributed by Raven Press, New York.

32 Robert Katzman's editorial, "The Prevalence and Malignancy of Alzheimer Disease; A Major Killer," appeared in *Archives of Neurology* 33, 217–218 (1976).

38 A healthy brain weighs, on average, approximately 1,300–1,400 grams, whereas one affected by Alzheimer's often weighs in the neighborhood of 1,000 grams.

40 Glenner's recalled conversation with President Reagan appeared in the *San Diego Union-Tribune* on April 6, 1995.

Studies

Alois Alzheimer's original case description of the disease:

Alzheimer, A. "Über eine eigenartige Erkrankung der Hirnrinde," *Allgemeine Zeitschrift für Psychiatrie und Psychisch-Gerichtliche Medizin* 64, 146–148 (1907).

Original description of the structural components of the tangles:

Kidd, M., "Paired Helical Filaments in Electron Microscopy of Alzheimer's Disease," *Nature* 197, 192–193 (1963).

Early study showing the structure of senile plaques:

Terry, R. D., Gonatas, N. K., Weiss M., "Ultrastructural Studies in Alzheimer's Presenile Dementia," *American Journal of Pathology* 44, 269–287 (1964).

Landmark report identifying Alzheimer's in older people:
Blessed, G., Tomlinson, B. E., and Roth, M., "The Association Between Quantitative Measures of Dementia and of Senile Change in the Cerebral Grey Matter of Elderly Subjects," *British Journal of Psychiatry* 114, 797–811 (1968).

Study showing that amyloid contains a beta-pleated sheet structure:
Eanes, E. D., and Glenner, G. G., "X-ray Diffraction Studies of Amyloid Filaments," *Journal of Histochemistry and Cytochemistry* 16, 673–677 (1968).

Representative study demonstrating cholinergic deficit in Alzheimer's:
Davies, P., and Maloney, A. J. F., "Selective Loss of Central Cholinergic Neurons in Alzheimer's Disease," *Lancet* 2, 1403–1400 (1976).

Glenner and Wong's initial report on the beta peptide:
Glenner, G. G., and Wong, C. W., "Alzheimer's Disease: Initial Report of the Purification and Characterization of a Novel Cerebrovascular Amyloid Protein," *Biochemical and Biophysical Research Communications* 120, 885–890 (1984).

Master and Beyreuther's paper on sequencing the beta peptide in plaque amyloid:
Masters, C. L., et al., "Amyloid Plaque Core Protein in Alzheimer Disease and Down Syndrome," *Proceedings of the National Academy of Sciences* (USA) 82, 4245–4249 (1985).

Chapter 3

57 The letter written by the British geneticists about Jim Gusella's withholding of the G8 probe, as well as Gusella's reply, appeared in *Nature* 320, 21–22 (1986).

Studies

First study to show the existence of infectious proteins called prions:

Prusiner, S. B., "Novel Proteinaceous Infectious Particles Cause Scrapie," *Science* 216, 136–144 (1982).

Report showing that brain blood-vessel amyloid is the same in Alzheimer and Down Syndrome patients: Glenner, G. G., and Wong, C. W., "Alzheimer's Disease and Down's Syndrome: Sharing of a Unique Cerebrovascular Amyloid Fibril Protein," *Biochemical and Biophysical Research Communications* 122, 1131–1135 (1984).

Chapter 4

66 The German Alzheimer family was assembled by Dr. Peter Frommelt at the Bavaria Clinic in West Germany. The Russian family was collected largely by Daniel Pollen, then at the University of Massachusetts Medical School, with the assistance of David Drachman at the same institution.

67 Glial cells are thought to constitute over 80 percent of brain cells. In the past, they've received far less attention than neurons. Yet today, because of new signs that they might be involved in memory, they're attracting more vigorous research.

78 Figures for the informal study of women attending the Society for Neuroscience meetings were provided by the Society for Neuroscience.

81 Goldgaber, asked about his recollections of that November day in Washington in 1986, maintains that I requested his lab's clone for the sake of doing genetic studies with the Mass General families and did not bring evidence of our lab's amyloid gene to his lab.

Studies

Paper describing the insolubility of tangles:
Selkoe, D. J., Ihara, Y., and Salazar, F. J., "Alzheimer's Disease: Insolubility of Partially Purified Paired Helical Filaments in Sodium Dodecyl Sulfate and Urea," *Science* 215, 1243–1245 (1982).

Isolation of the muscular dystrophy gene:
Monaco, A. P., et al., "Isolation of Candidate cDNAs for Portions of the Duchenne Muscular Dystrophy Gene," *Nature* 323, 646–650 (1986).

Four papers describing the isolation of the amyloid precursor protein (APP) and the localization of its gene to chromosome 21:

Goldgaber, D., Lerman, M. I., McBride, O. W., Saffiotti, U., and Gajdusek, D. C., "Characterization and Chromosomal Localization of a cDNA Encoding Brain Amyloid of Alzheimer's Disease," *Science* 235, 877–880 (1987).

Kang, J., et al., "The Precursor of Alzheimer's Disease Amyloid A4 Protein Resembles a Cell-surface Receptor," *Nature* 325, 733–736 (1987).

Robakis, N. K., et al., "Chromosome 21q21 Sublocalisation of Gene Encoding ß-amyloid Peptide in Cerebral Vessels and Neuritic (Senile) Plaques of People with Alzheimer Disease and Down Syndrome," *Lancet* 1, 384–385 (1987).

Tanzi, R. E., et al., "Amyloid ß Protein Gene: cDNA, mRNA Distribution, and Genetic Linkage Near the Alzheimer Locus," *Science* 235, 880–884 (1987).

Chapter 5

90 The triplication theory remains largely unsupported, although molecular biologist Huntington Potter—a professor of mine at Harvard now based at the University of South Florida—believes that a small number of cells might be undergoing division in the embryonic brain, which results in a triplication of chromosome 21 that progressively, over a person's lifetime, produces higher levels of the A-beta peptide and causes Alzheimer's. His pursuit of fuller evidence continues.

Studies

Miriam Schweber's hypothesis:

Schweber, M. A., "Possible Unitary Genetic Hypothesis for Alzheimer's Disease and Down Syndrome," *Annuals of the New York Academy of Sciences* 450, 223–238 (1985).

Study showing linkage of four familial Alzheimer families to chromosome 21:

St. George-Hyslop, P. H., et al., "The Genetic Defect Causing Familial Alzheimer's Disease Maps on Chromosome 21," *Science* 235, 885–890 (1987).

Two studies in which the APP gene was shown not to be linked to familial Alzheimer's disease:

Tanzi, R. E., et al., "The Genetic Defect in Familial Alzheimer's Disease is Not Tightly Linked to the Amyloid ß-protein Gene," *Nature* 329, 156–157 (1987).

Van Broeckhoven, C., et al., "Failure of Familial Alzheimer's Disease to Segregate with the A4-amyloid Gene in Several European Families," *Nature* 329, 153–155 (1987).

Study suggesting that the APP gene might be duplicated in Alzheimer patients:

Delabar, J. M., et al., ß Amyloid Gene Duplication in Alzheimer's Disease and Karyotypically Normal Down Syndrome," *Science* 235, 1390–1392 (1987).

Three papers showing that APP is not duplicated in Alzheimer's disease patients:

Tanzi, R. E., Bird, E. D., Latt, S. A., and Neve, R. L., "The Amyloid ß Protein Gene is Not Duplicated in Brains from Patients with Alzheimer's Disease," *Science* 238, 666–669 (1987).

St. George-Hyslop, P. H., et al., "Absence of Duplication of Chromosome 21 Genes in Familial and Sporadic Alzheimer's Disease," *Science* 238, 664–666 (1987).

Podlisny, M. B., Lee, G., and Selkoe, D. J., "Gene Dosage of the Amyloid ß Precursor Protein in Alzheimer's Disease," *Science* 238, 669–671 (1987).

Three papers describing the protease inhibitor insert in APP:

Tanzi, R. E., et al., "Protease Inhibitor Domain Encoded by an Amyloid Protein Precursor mRNA Associated with Alzheimer's Disease," *Nature* 331, 528–530 (1988).

Kitaguchi, N., Takahashi, Y., Tokushima, Y., Shiojiri, S., and Ito, H., "Novel Precursor of Alzheimer's Disease Amyloid Protein Shows Protease Inhibitory Activity," *Nature* 331, 530–532 (1988).

Ponte, P., et al., "A New A4 Amyloid mRNA Contains a Domain Homologous to Serine Proteinase Inhibitors," *Nature* 331, 525–527 (1988).

Papers collectively showing that the protease inhibitor form of APP is actually protease nexin II, a coagulation pathway protein:

Van Nostrand, W. E., and Cunningham, D. D., "Purification of Protease Nexin II from Human Fibroblasts," *Journal of Biological Chemistry* 262, 8508–8514 (1987).

Van Nostrand, W. E., et al., "Protease Nexin-II, a Potent Antichymotrypsin, Shows Identity to Amyloid ß-protein Precursor," *Nature* 341, 546–549 (1989).

Oltersdorf, T., et al., "The Secreted Form of the Alzheimer's Amyloid Precursor Protein with the Kunitz Domain is Protease Nexin-II," *Nature* 341, 144–147 (1989).

Chapter 6

106 Harvard still hands out diplomas made of parchment.

Studies

Representative papers showing that tangles contain the protein called tau:

Nukina, N., and Ihara, Y., "One of the Antigenic Determinants of Paired Helical Filaments is Related to Tau Protein," *Journal of Biochemistry* (Tokyo) 99, 1541–1544 (1986).

Grundke-Iqbal, I., et al., "Microtubule-associated Protein Tau. A Component of Alzheimer Paired Helical Filaments," *Journal of Biological Chemistry* 261, 6084–6089 (1986).

Kosik, K. S., Joachim, C. L., and Selkoe, D. J., "Microtubule-associated Protein Tau (Tau) is a Major Antigenic Component of Paired Helical Filaments in Alzheimer Disease," *Proceedings of the National Academy of Sciences* (USA) 83, 4044–4048 (1986).

Biochemical evidence that the cores of tangles are made up of tau:

Wischik, C. M., et al., "Isolation of a Fragment of Tau Derived from the Core of the Paired Helical Filament of Alzheimer Disease," *Proceedings of the National Academy of Sciences* (USA) 85, 4884–4888 (1988).

Goedert, M., Wischik, C. M., Crowther, R. A., Walker, J. E., and Klug, A., "Cloning and Sequencing of the cDNA Encoding a Core Protein of the Paired Helical Filament of Alzheimer Disease: Identification as the Microtubule-associated Protein Tau," *Proceedings of the National Academy of Sciences* (USA) 85, 4051–4055 (1988).

Report of a fragment of APP that is toxic to nerve cells:

Yankner, B. A., et al., "Neurotoxicity of a Fragment of the Amyloid Precursor Associated with Alzheimer's Disease," *Science* 245, 417–420 (1989).

Two reports showing that normally a protease cleaves APP within the A-beta region preventing amyloid formation:

Sisodia, S. S., Koo, E. H., Beyreuther, K., Unterbeck, A., and Price, D. L., "Evidence that ß-amyloid Protein in Alzheimer's Disease is not Derived by Normal Processing," *Science* 248, 492–495 (1990).

Esch, F. S., et al., "Cleavage of Amyloid ß Peptide During Constitutive Processing of its Precursor," *Science* 248, 1122–1124 (1990).

First report of an amyloid in the brain:

Cohen, D. H., Feiner, H., Jensson, O., and Frangione, B., "Amyloid Fibril in Hereditary Cerebral Hemorrhage with Amyloidosis (HCHWA) is Related to the Gastroentero-pancreatic Neuroendocrine Protein, Gamma Trace," *Journal of Experimental Medicine* 158, 623–628 (1983).

Chapter 7

Notes

121 In regard to the chromosome 19 data, previous data from Jerry Schellenberg's lab in 1987 had correlated a gene on chromosome 19 with Alzheimer's disease: Schellenberg, G. D., et al., "Association of an Apolipoprotein CII Allele with Familial Dementia of the Alzheimer Type," *Journal of Neurogenetics* 4, 97–108 (1987).

Studies

Reports of the APP gene mutation associated with amyloid formation in the blood vessels of patients from a Dutch family with a rare amyloidosis:

Levy, E., et al., "Mutation of the Alzheimer's Disease Amyloid Gene in Hereditary Cerebral Hemorrhage, Dutch Type," *Science* 248, 1124–1126 (1990).

Van Broeckhoven, C., et al., "Amyloid ß Protein Precursor Gene and Hereditary Cerebral Hemorrhage with Amyloidosis (Dutch)," *Science* 248, 1120–1122 (1990).

Report of the first Alzheimer gene mutation (in APP):

Goate, A., et al., "Segregation of a Missense Mutation in the Amyloid Precursor Protein Gene with Familial Alzheimer's Disease," *Nature* 349, 704–706 (1991).

Study showing linkage of late-onset Alzheimer's to chromosome 19:

Pericak-Vance, M. A., et al., "Linkage Studies in Familial Alzheimer Disease: Evidence for Chromosome 19 Linkage," *American Journal of Human Genetics* 48, 1034–1050 (1991).

Studies showing direct toxicity of A-beta on nerve cells in culture and rat brain:

Yankner, B. A., Duffy, L. K., and Kirschner, D. A., "Neurotrophic and Neurotoxic Effects of Amyloid ß Protein: Reversal by Tachykinin Neuropeptides," *Science* 250, 279–282 (1990).

Kowall, N. W., Beal, M. F., Busciglio, J., Duffy, L. K., and Yankner, B. A., "An In Vivo Model for the Neurodegenerative Effects of ß Amyloid and Protection by Substance P," *Proceedings of the National Academy of Sciences* (USA) 88, 7247–7251 (1991).

Demonstration that the toxicity of A-beta requires aggregation of the peptide into amyloid fibrils achieved by simply aging the peptides in a test tube:

Pike, C. J., Walencewicz, A. J., Glabe, C. G., and Cotman, C. W., "In Vitro Aging of ß-amyloid Protein Causes Peptide Aggregation and Neurotoxicity," *Brain Research* 563, 311–314 (1991).

Reports of two more Alzheimer mutations in APP at the same site as the one originally described by Goate et al. in the same year:

Murrell, J., Farlow, M., Ghetti, B. and Benson, M. D., "A Mutation in the Amyloid Precursor Protein Associated with Hereditary Alzheimer's Disease," *Science* 254, 97–99 (1991).

Chartier-Harlin, M. C., et al., "Early-onset Alzheimer's Disease Caused by Mutations at Codon 717 of the ß-amyloid Precursor Protein Gene," *Nature* 353, 844–846 (1991).

Chapter 8

Notes

135 A good indicator of the terrific growth in Alzheimer research is Medline. The medical journal database reveals that the number of English-language papers with "Alzheimer's Disease" as a subject heading surged from 95 in 1980, to 674 in 1987, to 1130 in 1991.

142 Lewis Thomas's essay "On the Problem of Dementia," which includes his "disease-of-the-century" quote, first ran in *Discover*'s August 1981 issue.

143 The authors have it on good authority that the terms "Tauist" and "Baptist" first appeared in a journal editorial, but we have been unable to pin down the exact source. The terms were in use by early 1991.

Studies

Study showing that the tangles are made up of tau that has been modified by large amounts of bound phosphate:

Lee, V. M. Y., Balin, B. J., Otvos, L., Jr., and Trojanowski, J. Q., "A68: A Major Subunit of Paired Helical Filaments and Derivatized Forms of Normal Tau," *Science* 251, 675–678 (1991).

Early transgenic mouse model that fell short of full-blown Alzheimer's disease pathology:

Quon, D., Wang, Y., Catalano, R., Scardina, J. M., Murakami, K., and Cordell, B., "Formation of Beta-amyloid Protein Deposits in Brains of Transgenic Mice," *Nature* 352, 239–41 (1991)

Study showing a prevalence of Alzheimer's in the general population that was much higher than previously assumed. Close to 50 percent of individuals over 85 years in the Boston sample were found to have dementia of the Alzheimer type:

Evans, D. A., et al., "Prevalence of Alzheimer's Disease in a Community Population of Older Persons," *Journal of the American Medical Association* 262, 2551–2556 (1989).

Chapter 9

Studies

Report showing that amyloid plaques contain alpha1-antichymotrypsin, the gene for which resides on chromosome 14:

Abraham, C. R., Selkoe, D. J., and Potter, H., "Immunochemical Identification of the Serine Protease Inhibitor Alpha 1-antichymotrypsin in the Brain Amyloid Deposits of Alzheimer's Disease," *Cell* 52, 487–501 (1988).

First report to show linkage of early-onset familial Alzheimer's disease to chromosome 14:

Schellenberg, G. D., et al., "Genetic Linkage Evidence for a Familial Alzheimer's Disease Locus on Chromosome 14," *Science* 258, 668–671 (1992).

Three papers confirming the linkage of chromosome 14 to early-onset familial Alzheimer's disease:

Mullan, M., et al., "A Locus for Familial Early-onset Alzheimer's Disease on the Long Arm of Chromosome 14, Proximal to the Alpha 1-antichymotrypsin Gene," *Nature Genetics* 2, 340–342 (1992).

St. George-Hyslop, P., et al., "Genetic Evidence for a Novel Familial Alzheimer's Disease Locus on Chromosome 14," *Nature Genetics* 2, 330–334 (1992).

Van Broeckhoven, C., et al., "Mapping of a Gene Predisposing to Early-onset Alzheimer's Disease to Chromosome 14q24.3," *Nature Genetics* 2, 335–339 (1992).

Report showing that A-beta binds to the protein apolipoprotein E. In the same paper, a form of apolipoprotein E called APOE-4 is found to be genetically associated with late-onset Alzheimer's:

Strittmatter, W. J., et al., "Apolipoprotein E: High-avidity Binding to ß-amyloid and Increased Frequency of Type 4 Allele in Late-onset Familial Alzheimer Disease," *Proceedings of the National Academy of Sciences* (USA) 90, 1977–1981 (1993).

One of many confirmatory reports showing that the APOE gene is a risk factor for late-onset Alzheimer's disease in sporadic cases and families:

Corder, E. H., et al., "Gene Dose of Apolipoprotein E Type 4 Allele and the Risk of Alzheimer's Disease in Late Onset Families," *Science* 261, 921–923 (1993).

Representative papers first showing that A-beta is normally secreted by nerve cells and other types of cells in the body and is found in bodily fluids:

Seubert, P., et al., "Isolation and Quantification of Soluble Alzheimer's ß-peptide from Biological Fluids," *Nature* 359, 325–327 (1992).

Haass, C., et al., "Amyloid ß-peptide is Produced by Cultured Cells During Normal Metabolism," *Nature* 359, 322–325 (1992).

Shoji, M., et al., "Production of the Alzheimer Amyloid ß Protein by Normal Proteolytic Processing," *Science* 258, 126–129 (1992).

Busciglio, J., Gabuzda, D. H., Matsudaira, P., and Yankner, B. A., "Generation of ß-amyloid in the Secretory Pathway in Neuronal and Nonneuronal Cells," *Proceedings of the National Academy of Sciences* (USA) 90, 2092–2096 (1993).

Studies showing that an Alzheimer mutation in the APP gene causes excess production of A-beta:

Citron, M., et al., "Mutation of the ß-amyloid Precursor Protein in Familial Alzheimer's Disease Increases ß-protein Production," *Nature* 360, 672–674 (1992).

Seubert, P., et al., "Secretion of ß-amyloid Precursor Protein Cleaved at the Amino Terminus of the ß-amyloid Peptide," *Nature* 361, 260–263 (1993).

Identification of the Huntington's disease gene:

The Huntington's Disease Collaborative Research Group. "A Novel Gene Containing a Trinucleotide Repeat that is Expanded and Unstable on Huntington's Disease Chromosomes," *Cell* 72, 971–983 (1993).

Report of the fourth Alzheimer mutation in APP, this time in a Swedish family:

Mullan, M., et al., "A Pathogenic Mutation for Probable Alzheimer's Disease in the APP Gene at the N-terminus of ß-amyloid," *Nature Genetics* 1, 345–347 (1992).

First report showing that some Alzheimer mutations in APP lead to an increased percentage of a longer form of A-beta called A-beta 42 (the more prevalent form being A-beta 40):

Suzuki, N., et al., "An Increased Percentage of Long Amyloid ß Protein Secreted by Familial Amyloid ß Protein Precursor (ß APP717) Mutants," *Science* 264, 1336–1340 (1994).

Study showing that the longer form of A-beta, A-beta 42, seeds the formation of amyloid plaques and is thus the "bad" A-beta:

Jarrett, J. T., Berger, E. P., and Lansbury P.T, Jr., "The Carboxy Terminus of the ß Amyloid Protein is Critical for the Seeding of Amyloid Formation: Implications for the Pathogenesis of Alzheimer's Disease," *Biochemistry* 32, 4693–4697 (1993).

Initial identification of presenilin 1, the early-onset familial Alzheimer's disease gene on chromosome 14:

Sherrington, R., et al., "Cloning of a Gene Bearing Missense Mutations in Early-onset Familial Alzheimer's Disease," *Nature* 375, 754–760 (1995).

Chapter 10

Notes

181 To date, over 650 aides have received training at the Glenner School of Dementia Care.

Studies

Initial identification of presenilin 2, the early-onset familial Alzheimer gene on chromosome 1:

Levy-Lahad, E., et al., "Candidate Gene for the Chromosome 1 Familial Alzheimer's Disease Locus," *Science* 269, 973–977 (1995).

Two confirmatory reports of the presenilin 2 gene on chromosome 1:

Rogaev, E. I., et al., Familial Alzheimer's Disease in Kindreds with Missense Mutations in a Gene on Chromosome 1 Related to the Alzheimers Disease Type 3 Gene," *Nature* 376, 775–778 (1995).

Li, J. H., Ma, J. L., and Potter, H., "Identification and Expression Analysis of a Potential Familial Alzheimer Disease Gene on Chromosome 1 Related to AD3," *Proceedings of the National Academy of Sciences* (USA) 92, 12180–12184 (1995).

Study showing that patients with presenilin mutations make more A-beta 42 as evidenced in blood plasma and skin cells:

Scheuner, D., et al., "Secreted Amyloid ß-protein Similar to that in the Senile Plaques of Alzheimer's Disease is Increased In Vivo by the Presenilin 1 and 2 and APP Mutations Linked to Familial Alzheimer's Disease," *Nature Medicine* 2, 864–870 (1996).

First study to show where presenilins are localized in cells:

Kovacs, D. M., et al., "Alzheimer-associated Presenilins 1 and 2: Neuronal Expression in Brain and Localization to Intracellular Membranes in Mammalian Cells," *Nature Medicine* 2, 224–229 (1996).

Reports of the first successful transgenic mice exhibiting Alzheimer type features:

Games, D., et al., "Alzheimer-type Neuropathology in Transgenic Mice Overexpressing V717F ß-amyloid Precursor Protein," *Nature* 373, 523–527 (1995).

Hsiao, K., et al., "Correlative Memory Deficits, Aß Elevation, and Amyloid Plaques in Transgenic Mice," *Science* 274, 99–102 (1996).

Duff, K., et al., "Increased Amyloid-ß42(43) in Brains of Mice Expressing Mutant Presenilin 1," *Nature* 383, 710–713 (1996).

Study showing that A-beta begins to accumulate in the brains of fetuses with Down Syndrome:

Teller, J. K., et al., "Presence of Soluble Amyloid Beta-Peptide Precedes Amyloid Plaque Formation in Down's Syndrome," *Nature Medicine* 2, 93-95 (1996).

Chapter 11

Notes

197 The Eli Lilly study showing a correlation between high cholesterol and brain amyloid was presented at the International Business Communications' annual conference on Alzheimer's disease in May 1999.

198 A study appearing in the April 2000 issue of the *Journal of the American College of Nutrition* reports that middle-aged men who consumed tofu two or more times a week showed more evidence of mental de-

cline than those who consumed less tofu. A possible explanation is that phyto-estrogens in tofu may be blocking cell activities that contribute to learning.

Studies

Study of Alzheimer's in identical twins suggesting a role for both genetic and environmental factors in the disease:

Breitner, J. C., Murphy, E. A., Folstein, M. F., and Magruder-Habib, K., "Twin Studies of Alzheimer's Disease: An Approach to Etiology and Prevention," *Neurobiology of Aging* 11, 641–648 (1990).

Paper showing that zinc can induce the rapid formation of beta amyloid when mixed with A-beta in a test tube:

Bush, A. I., et al., "Rapid Induction of Alzheimer Aß Amyloid Formation by Zinc," *Science* 265, 1464–1467 (1994).

Twin study suggesting that anti-inflammatory agents may protect against Alzheimer's disease:

Breitner, J. C., et al., "Inverse Association of Anti-inflammatory Treatments and Alzheimer's Disease: Initial Results of a Co-twin Control Study," *Neurology* 44, 227–232 (1994).

Review of the role of inflammation in the brains of Alzheimer patients and summary of a recent trial of an inflammatory drug:

McGeer, E. G., and McGeer, P. L., "The Importance of Inflammatory Mechanisms in Alzheimer Disease," *Experimental Gerontology* 33, 371–378 (1998).

Study showing that vitamin E together with selegiline slows progress of Alzheimer's disease:

Sano, M., et al., "A Controlled Trial of Selegiline, Alpha-tocopherol, or Both as Treatment for Alzheimer's Disease," The Alzheimer's Disease Cooperative Study, *New England Journal of Medicine* 336, 1216–1222 (1997).

Representative studies showing that Alzheimer mutations in the presenilins may predispose to programmed cell death:

Wolozin, B., et al., "Participation of presenilin 2 in Apoptosis: Enhanced Basal Activity Conferred by an Alzheimer Mutation," *Science* 274, 1710–1713 (1996).

Kovacs, D. M., et al., "Staurosporine-induced Activation of Caspase-3 is Potentiated by Presenilin 1 Familial Alzheimer's Disease Mutations in Human Neuroglioma Cells," *Journal of Neurochemistry* 73, 2278–2285 (1999).

Epidemiological study suggesting estrogen might protect against Alzheimer's:
Steffens, D. C., et al., "Enhanced Cognitive Performance with Estrogen Use in Nondemented Community-dwelling Older Women," *American Geriatric Society* 47, 1171–1175 (1999).

Recent trial showing that estrogen was not effective against mild to moderate Alzheimer's disease:
Mulnard, R. A., et al., "Estrogen Replacement Therapy for Treatment of Mild to Moderate Alzheimer Disease: A Randomized Controlled Trial," Alzheimer's Disease Cooperative Study, *Journal of the American Medical Association* 283, 1007–1015 (2000).

Evidence that boxers who carry the APOE-4 gene are more vulnerable to dementia due to head trauma:
Jordan, B. D., Relkin, N. R., Ravdin, L. D., Jacobs, A. R., Bennett, A., Gandy, S., "Apolipoprotein E Epsilon4 Associated with Chronic Traumatic Brain Injury in Boxing," *Journal of the American Medical Association* 278, 136-140 (1997).

Demonstration that A-beta's generation is lessened by the administration of testosterone:
Gouras, G. K., et al., "Testosterone Reduces Neuronal Secretion of Alzheimer's Beta-Amyloid Peptides," *Proceedings of the National Academy of Sciences (USA)* 97, 1202-1205 (2000).

Chapter 12

Notes

230 An observation by Nan Laird: "It isn't widely appreciated, but statistical methodology serves as a sine qua non in this era of linking genes to human disease. The new statistical approach that led to A2M's linkage to Alzheimer's is a prime illustration of this often neglected, statistical side of the gene-hunt story. When all human genes are mapped, we'll need even more advances in statistical methodology if we're going to figure out the complex role that thousands of candidate risk-factor genes—susceptibility genes just like A2M—play in human diseases."

Studies

The Duke scientists' chromosome 12 linkage paper:

Pericak-Vance, M. A., et al., "Complete Genomic Screen in Late-onset Familial Alzheimer Disease. Evidence for a New Locus on Chromosome 12," *Journal of the American Medical* Association 278, 1237–1241 (1997).

Representative papers reporting mutations in the tau gene that cause frontal-temporal dementia:

Hutton, M., et al., "Association of Missense and 5'-splice-site Mutations in Tau with the Inherited Dementia FTDP–17," *Nature* 393, 702–705 (1998).

Poorkaj, P., et al., "Tau is a Candidate Gene for Chromosome 17 Frontotemporal Dementia," *Annals of Neurology* 43, 815–825 (1998).

Spillantini, M. G., et al., "Mutation in the Tau Gene in Familial Multiple System Tauopathy with Presenile Dementia," *Proceeding of the National Academy of Sciences* (USA) 95, 7737–7741 (1998).

Report showing that risk for late-onset Alzheimer's disease is associated with the A2M gene:

Blacker, D., et al., "Alpha–2 Macroglobulin is Genetically Associated with Alzheimer Disease," *Nature Genetics* 19, 357–360 (1998).

Dennis Selkoe's lab finding that presenilin could be the gamma-secretase:

Wolfe, M. S., et al., "Two Transmembrane Aspartates in Presenilin–1 Required for Presenilin Endoproteolysis and Gamma-secretase Activity," *Nature* 398, 513–517 (1999).

Vaccine for Alzheimer's disease:

Schenk, D., et al., "Immunization with Amyloid-beta Attenuates Alzheimer-disease-like Pathology in the PDAPP Mouse," *Nature* 400, 173–177 (1999).

First beta-secretase report:

Vassar, R., et al., "Beta-Secretase Cleavage of Alzheimer's Amyloid Precursor Protein by the Transmembrane Aspartic Protease BACE," *Science* 286, 735–741 (1999).

Epilogue

Studies

Paper suggesting that there are five to six major Alzheimer genes responsible for the late-onset form in addition to the APOE gene:

Warwick, Daw E., et al., "The Number of Trait Loci in Late-Onset Alzheimer Disease," *American Journal of Human Genetics* 66, 196-204 (2000).

Report of a mutation that leads to a form of brain amyloid similar to beta-amyloid:

Vidal, R., et al., "A Stop-Codon Mutation in the BRI Gene Associated with Familial British Dementia," *Nature* 399, 776-81 (1999).

Study showing that A-beta peptide can trigger an inflammatory response in the brain by activating microglia:

Tan, J., et al., "Microglial Activation Resulting from CD40-CD40L Interaction after Beta-Amyloid Stimulation," *Science* 286, 2352-5 (1999).

Study demonstrating that the toxicity of A-beta is greater in the brains of old verus young primates or young mice:

Geula, C., Wu, C. K., Saroff, D., Lorenzo, A., Yuan, M., Yankner, B. A., "Aging Renders the Brain Vulnerable to Amyloid Beta-Protein Neurotoxicity," *Nature Medicine* 4, 827-31 (1998).

Report showing amounts of brain A-beta parallel degree of dementia:

Näslund, J., et al., "Correlation Between Elevated Levels of Amyloid Beta-Peptide in the Brain and Cognitive Decline," *Journal of the American Medical Association* 283, 1571-7 (2000).

Report showing that structural changes in the entorhinal cortex (involved with sensory memory) strongly correlate with the advancement to full-blown Alzheimer's disease:

Killany, R.J., et al., "Use of Structural Magnetic Resonance Imaging to Predict Who Will Get Alzheimer's Disease," *Annals of Neurology* 47, 430-439 (2000).

Index